BOTANICAL EN

Botanical Entanglements

Women, Natural Science, and the Arts in Eighteenth-Century England

Anna K. Sagal

University of Virginia Press
Charlottesville and London

University of Virginia Press

Printed in the United States of America on acid-free paper

First published 2022

1 3 5 7 9 8 6 4 2

Library of Congress Cataloging-in-Publication Data

Names: Sagal, Anna K., author.
Title: Botanical entanglements : women, natural science, and the arts in eighteenth-century England / Anna K. Sagal.
Description: Charlottesville : University of Virginia Press, 2022. | Includes bibliographical references and index.
Identifiers: LCCN 2021030591 (print) | LCCN 2021030592 (ebook) | ISBN 9780813946955 (hardcover) | ISBN 9780813946962 (paperback) | ISBN 9780813946979 (ebook)
Subjects: LCSH: Natural history—England—History—18th century. | Women naturalists—England—History—18th century. | Art and natural history—England—History—18th century. | Natural history literature—Women authors.
Classification: LCC QH21.G7 S24 2021 (print) | LCC QH21.G7 (ebook) | DDC 508.4109/033—dc23
LC record available at https://lccn.loc.gov/2021030591
LC ebook record available at https://lccn.loc.gov/2021030592

Cover art: Top, "Indian Chrysanthemum," Henrietta Maria Moriarty, 1806; *middle,* "Philadelphian Lily," Henrietta Maria Moriarty, 1807; *bottom,* "Fig Tree," Elizabeth Blackwell, 1737. (Top and middle images from the Rare Book Collection of the Lenhardt Library of the Chicago Botanic Garden; bottom image from the Biodiversity Heritage Library; contributed by Missouri Botanical Garden, Peter H. Raven Library)

CONTENTS

ACKNOWLEDGMENTS

THE BOOK you're holding in your hands (or reading on your screen) would not have been possible without the generous support of a number of institutions and the even more generous support of friends and colleagues. The McMaster Archives and Research Collections was an especially formative place for many of the ideas about women's reconceptualizing of human/plant relationships. Thanks to Bridget Whittle and Myron Groover for their invaluable research assistance and enthusiastic recommendations for botanically inclusive romance novels and local craft beer. And, although our relationships were not tied to this fellowship, I valued the opportunity to spend more time with Gena Zuroski and Jacqueline Langille, whose support of early versions of chapter 1 in *Eighteenth-Century Fiction* strengthened my confidence in this project. A fellowship at the Folger Shakespeare Library allowed me to spend several months poring over a vast array of periodicals and conduct books, most of which underpinned significant thinking about the chapters on pedagogy and didactic texts. Many thanks to Abbie Weinberg for all of her help. The Monticello College Foundation and Audrey Lumsden-Kouvel Fellowship at the Newberry Library provided priceless time, resources, and space to fully develop the project. I spent more time looking at illustrated botanical volumes in the reading rooms than was strictly necessary, but it's hard to turn down the chance to peruse *Curtis's Botanical Magazine*, even if you have nothing more to say about it in writing. Thanks to Brad Hunt for the support, Keelin Burke for the camaraderie, and Jill Gage for advice and ideas. The Rare Book and Manuscript Library at the University of Illinois at Urbana-Champaign also, very conveniently, had a copy of Henrietta Maria Moriarty's *Brighton in an Uproar* when I decided I needed to very urgently consult it one summer. Thanks to Lynne Thomas for her support the subsequent summer, at which point I was more methodical about the resources I spent time with as a Visiting Scholar. I also very much appreciated the chance to sit with Moriarty's *Fifty Plates of Green-House Plants* for hours at the Chicago Botanic Garden's Lenhardt Library, after which a stroll through the gardens seemed like the most perfect thing in the world. Lastly, the writing of this book would not have been possible without the tireless efforts of the librarians with whom I have been fortunate to work. For fulfilling a truly impressive (and probably unreasonable) number of interlibrary loan requests and tracking down even the most obscure writing on botany and critical plant theory, many, many thanks to Randi Thon. Likewise, my undying gratitude to Ellen Wrede for granting the most flexible due dates and renewal chances I've ever heard of, let alone benefitted from personally.

For reading very (very) early versions of some of this project, my sincere thanks to Andrea Haslanger, Carol Flynn, and Kevin Dunn. Beth Kowaleski-Wallace not only introduced me to Mary Delany, but also encouraged me to expand the theoretical framework of the book from its first life as an enthusiastic but unfocused project on women's scientific interests in the eighteenth century. During the year-long GCWS workshop at MIT, Kimberly Juanita Brown inspired me to vigorously embrace the feminist stakes of the project. Over the course of the workshop Nicole Keller became not only a writing partner but a good friend, and I continue to benefit from her support. Thanks to Danielle Spratt, Adela Ramos, and Maddy Pelling for their careful reading of various chapters—I value your insight and expertise and my scholarship is far better for it. Particular gratitude to Melissa Bailes, who played many roles in the success of this book: *Botanical Entanglements* would not be the same without you! And, of course, a profound thank-you to Kelly Fleming, who read just about everything (many times, over many years). To the anonymous reviewers for the press, I thank you for your incredibly generous feedback, brilliant insight, and helpful suggestions at multiple stages of the project. I was also fortunate enough to be part of a number of writing groups that offered emotional support and fellowship during the long, arduous process of writing. Sarah Creel and Mary Beth Harris, thank you for being my writing group/cheer squad during a difficult summer. Thanks also to Rebecca Shapiro and Emily Friedman, who invited me to join their daily virtual cowriting sessions in 2020. It was during these sessions, in the first few months of the pandemic, that much of the final work on the book was completed. Thank you to Nush Powell for your unwavering support of my Eliza Haywood obsession and for helping me to refine the ideas in chapter 1. I have also enjoyed the tireless support of Jess Keiser throughout the duration of this project at many different stages, for which I am eternally grateful. My colleagues in the English and Creative Writing Department at Cornell College have also been consistently excited about this project. Finally, my thanks to Angie

Hogan, who saw the potential in this project years ago and has helped bring it to a successful publication.

For their enthusiasm and encouragement throughout all stages of this project, I thank my family. It's been a long, strange trip, but it's a book! Last, but also first and foremost, thank you to MGW. I would not be here today without you and this book would not be what is without your unconditional support and love. Thank you for enduring my obsessive focus on the project during rare weekends together in Boston, my tendency to jump immediately into an hour of writing when something occurred to me regardless of our plans, and my insistence upon meandering around every botanical garden within driving distance (and you can drive pretty far). Thank you, for everything.

An earlier version of a portion of chapter 1, in significantly different form, appeared as "'Philosophy for the Ladies': Feminism, Pedagogy, and Natural Philosophy in Charlotte Lennox's *The Lady's Museum*" in *Eighteenth-Century Fiction* 28, no. 1 (2015): 139–66. I thank the University of Toronto Press for permission to reproduce that work here. An earlier version of a different portion of chapter 1 appeared as "From 'Pretty Emulation' to 'Fresh Discoveries': Entomological Observation and Female Empiricism in Eliza Haywood's *The Female Spectator*" in *Restoration: Studies in English Literary Culture, 1660–1700* 44, no. 1 (2020): 87–104. I also thank *Restoration* for their permission to reproduce parts of that material here.

BOTANICAL ENTANGLEMENTS

Introduction

Botanical Entanglements is a book about the subversive ways in which women made their mark on the scientific landscape of eighteenth-century England. I argue that women's participation in the development of the natural sciences relied upon their clever manipulation of the very societal and cultural restraints that have previously been read as preventing such participation. In other words, their rootedness in domesticity and motherhood, their inextricability from femininity and all its attendant associations with excessive sexuality, and their practical exclusion from professional opportunities at home and abroad, typically understood as delimiting factors, were in fact potentially generative vectors for intellectual development. With a specific focus on botany and, to a lesser extent, natural history, the varied case studies in this book highlight the way the domestic spaces of home and garden provided prolific opportunities for personal inspiration, emotional fulfillment, and meaningful intellectual labor. More radically, this book also makes visible the corporeal and psychological intimacy between women, plants, and small insects—an intimacy that prefigures the theorizing of plant intelligence and vegetal culture that has emerged in twenty-first-century scholarship. By looking to the concept of "entanglement," I note how the subjects of this volume evocatively rework the harmful, restrictive dynamic between women and nature that was circulated in eighteenth-century culture as inviolable truth. To deploy the quotidian spaces of domestic gardens, fields, and woodlands for intellectual and emotional fulfillment is to subvert preconceived notions (usually involving sexual license and immorality) and produce instead new alliances between women and the natural world. This book tells the story of how such radical retooling was accomplished.

The narrative of the emergence of modern science from early concepts of knowledge formation and inquiry was traditionally a phallocentric story, in which Francis Bacon's intellectual legacy was brought to fruition, with the Royal Society's foundation in 1660. As the first and now the oldest scientific society in the Western world, this London-based group of men focused initially on the physical sciences, including chemistry, physics, and anatomy, with a subsequent expansion into the natural sciences, such as entomology, botany, and zoology, as the eighteenth century wore on.[1] Due to their extensive efforts to position themselves as exclusive cultural and scientific authorities (including a curiously premature *History of the Royal Society* penned by Thomas Sprat in 1667), for some time historians unquestioningly replicated their carefully curated narrative of supremacy and ultimate authority rooted in male intellectual circles. It is only relatively recently that historians have begun to dispute this hegemonic legacy, with vital philosophical interventions including work by Carolyn Merchant, Donna Haraway, Sandra Harding, and Evelyn Fox Keller published in the last decades of the twentieth century.[2]

Along with this theorizing of the relationship of women to science as a practice and an intellectual discourse came a new focus on recovering the role that real women played in the history of scientific thought. Women like Margaret Cavendish, Caroline Hershel, and Sophie Cuvier in domestic relationships with men in positions of established scientific authority (as wives, sisters, or daughters) have in the last several decades had their own labors acknowledged alongside remarkable women like Lucy Hutchinson, Anne Conway, and Émilie du Châtelet, whose singular contributions to seventeenth- and eighteenth-century scientific and intellectual discourse have garnered critical attention. Likewise, scholarship by Ann Shteir, Londa Schiebinger, Barbara Gates, Patricia Phillips, and Marina Benjamin, among others, successfully curated an archive of women's scientific labor of the seventeenth through nineteenth centuries that included everything from botanical sketches to chemical experimentation, thereby laying the foundation for later generations of scholars who have since nuanced such vital recovery work by expanding our understanding of scientific labor and participation.[3]

By acknowledging the value of the amateur culture of natural sciences in the second half of the eighteenth century and the early decades of the nineteenth century, contemporary studies of women's contributions to science have illuminated new venues for women's naturalist labor. While expert communities of men were traditionally seen as the sole locus for scientific research and discovery, it is more accurate to say that naturalist knowledge was formed in dialogue with and via contribution from amateur individuals (many of whom were women) with an active interest in botany, zoology, and natural history.[4] For one, Beth Fowkes Tobin has showcased the significant contributions of the Duchess of Portland to the vast networks of scientific research and exploration in the later eighteenth century.[5] Additional studies have centralized women with less fame and fortune who nevertheless had lasting impacts on botany, zoology, entomology, and paleontology.[6] Moreover, by expanding our conceptualization of what constituted meaningful achievement and contribution to a broad spectrum of scientific inquiry in the eighteenth century, we can find new connections between curious individuals across gendered, social, and economic lines. Thus, domestic activities ranging from the curation of household medicinal recipes to the dyeing of fabrics can be productively read within the context of scientific development.[7] Recent innovative scholarship by Melissa Bailes and Karen Gevirtz has likewise showcased the substantive naturalist knowledge exhibited and deployed by women authors, encouraging us to read literary genres like poetry and novels as themselves valuable contributions to the circulation of scientific knowledge in the period.[8]

Botanical Entanglements utilizes these productive insights to illuminate new perspectives on the history of female participation in the natural sciences via literature and art. The selection of texts I examine emphasize the significant connections between domesticity and scientific knowledge (a characteristic of eighteenth-century naturalist inquiry) and the inventive, often unexpected ways that women leveraged such intersections to undertake profound intellectual labor. Offering a sustained study of didactic texts, periodicals, prose fiction, and poetry, four chapters reveal the meaningful contributions of female authors to the ongoing discourse of naturalist knowledge from the 1740s to the 1810s. Further, by building upon recent scholarship in the history of material culture and domestic craft in eighteenth-century England, two chapters refresh the critical approach to reading women's botanical artwork in the period.[9] Uniting these varied threads of inquiry is an interest in the fruitful possibilities attainable through a critical alliance between women and nature. Through their focus on organic actors commonly represented as diminutive, insignificant, or unimportant, the women of this project imbued their studies of caterpillars, dandelions, and meadow grasses with the revolutionary promise of further enlightenment.

This book presents the eighteenth-century development of various strategies for revising women's relationality to nature in three particular sections. The first of these, featuring periodicals by Eliza Haywood and Charlotte Lennox, botanical textbooks by Priscilla Wakefield and

Maria Jacson, and multiple works by Maria Edgeworth, focuses on the role didactic literature played in creating a space for women in scientific pedagogy. The second section concentrates on women's art inspired by the natural world—specifically, illustrated reference works by Elizabeth Blackwell and Henrietta Maria Moriarty and Mary Delany's paper mosaics and other naturalist projects—and showcases the way these women pushed the boundaries of botanical representation. The third of these sections, the book's coda, is a brief, detailed study of Charlotte Smith's poem "Flora," in which Smith's distinctive poetic style reveals a subversive portrait of human-plant entanglements.

The Naturalist as Teacher

Botanical Entanglements begins by examining texts with a pedagogical investment in scientific subjects. While the significant contribution of women to the development of pedagogical literature in the late eighteenth century has long been acknowledged, it is only more recently that female-authored works centralizing scientific pedagogy have been reevaluated as important entries in an ongoing discourse about naturalist knowledge, rather than imitative texts that simply regurgitated knowledge generated by male intellectuals. Unsurprisingly, the writers studied in this chapter have received uneven critical focus in modern scholarship. For example, Maria Edgeworth has been the subject of much excellent scholarship critiquing and assessing her pedagogical bona fides, but attention to her scientific interests insofar as they intersect with her focus on female readers has been limited.[10] Priscilla Wakefield and Maria Jacson, on the other hand, have rarely received sustained attention to the extent that this project undertakes.[11] And, finally, while Eliza Haywood and Charlotte Lennox have become beloved topics of feminist literary scholarship in the last few decades, again, investment in exploring their scientific pedagogy is rare.[12]

In these three chapters, I evaluate how the assumption of a teaching role combined an expectation of maternal responsibility with a careful hedging of intellectual expertise to enable female writers to present themselves as scientific authorities. By conceptually and narratively positioning their expertise as teachers of naturalist subjects within the very organic space of nature itself, these authors contributed to what would later emerge as a sustained tradition of female pedagogy with a scientific emphasis.[13] Further, with a discrete focus on specific types of biota (including tiny unremarkable insects, common indigenous flora, and popular acclimatized plants), these women not only carved out space for themselves in naturalist discourses but did so in a way that meaningfully reshaped the significance of their subject matter.

Didactic texts such as textbooks, periodicals, essays, and prose fiction were not especially daring genres of writing for women to produce (although the periodical market was notoriously volatile). Yet the preponderance of pedagogical texts with a botanical focus were authored by men, with some of the most popular volumes on plants penned by everyone from nurseryman James Lee to polymath William Withering to novelist and philosopher Jean-Jacques Rousseau. Even as the "botanomania" of the second half of the eighteenth century generated enough enthusiasm for the subject that female authors had opportunities to contribute to the deluge of botanical publications, it was still challenging for them to be regarded as serious authorities on the subject. Women's involvement in entomological study was even more limited, although the subject was usually considered appropriate for women, given the ubiquity of their subjects and the low risk involved. Still, the question of intellectual credentials and scientific authority remained. In a strategic response to such doubts, a few of the authors in this section leveraged close relationships with male intellectual authorities in order to market their own scientific works; all of them also astutely linked their own scientific authority to the expertise of these male authors (citing their works in notes or referencing their ideas in-text). Simultaneously, these women also asserted their own independent conclusions and unique interpretations of the significance of naturalist study to the lived experience of being a woman in the eighteenth century.

Haywood and Lennox were not primarily concerned with scientific education in their popular periodicals, the *Female Spectator* and the *Lady's Museum*, but the parts of each publication that do explore naturalist study are remarkable for their distinctive pedagogical strategies. By providing limited but not diluted scientific content in their periodicals, both authors were able to claim scientific authority in conjunction with an already accepted presentation of authority on subjects like courtship, fashion, and female friendship. In chapter 1, I look in particular at their articulation of strategies for naturalist education (enabled specifically via the genre of the periodical) and their simultaneous cultivation of interspecies sympathies. By conceptually aligning themselves with their diminutive subjects, Haywood and Lennox reinscribe scientific value onto domestic vignettes of naturalist inquiry. Therefore, while expert scientific communities had shifted their investigations to exotic plants and insects discovered through colonial bioprospecting and exploitation, these women were able to claim domestic sites as invaluable sources of enlightenment and inspiration for curious female readers. The *Female Spectator* and the *Lady's Museum* both offered realistic strategies for women across economic and social spectrums to pursue independent naturalist inquiry through distinctive approaches to the dissemination of entomological information. Finally, in locating their passionate endorsement of naturalist study in the genre of the popular periodical, Haywood and Lennox also reminded their readers that pleasure is a viable (and valuable) component of scientific inquiry.

Although most texts that used the garden space as a site for teaching children usually expanded that focus to include moral and social education (such as Charlotte Smith's *Rural Walks* and *Rambles Farther,* as well as Sarah Trimmer's *An Easy Introduction to the Knowledge of Nature*), the textbooks discussed in chapter 2 almost exclusively concentrated on encouraging their readers to pursue botanical study. In genre, both Wakefield's *An Introduction to Botany* and Jacson's *Botanical Dialogues* employed formats with strong educational roots (the epistolary collection and dialogue, respectively), which enabled the authors to present themselves as sources of vital information on key subjects. In content, however, these texts were unusually rigorous and specific—particularly Jacson's—and the level of botanical detail was more impressive than the information presented in many contemporary texts aimed at female audiences. Both texts also actively encouraged the pursuit of naturalist inquiry outside the boundaries of the text: the readers were to use their newly acquired information to engage in the active practice of botanizing in their gardens or nearby green spaces. Acknowledging the profound significance of hands-on learning, Jacson and Wakefield facilitated their readers' endeavors by supplying them with information, advice, and resourceful tips, such as which common household tools could be adapted to botanical study. Finally, these texts touch upon questions of plant intelligence, compellingly articulating a continuity between vegetal bodies and human bodies that reveal contemporary tensions regarding the threat of vegetal monstrosity.

The significance of female networks to the successful pursuit of naturalist inquiry implied in the previous two chapters is made tangible in chapter 3. With an expansive focus on an array of Edgeworth's texts—ranging from *Letters for Literary Ladies* and excerpts from *Practical Education* to select *Moral Tales* and *Belinda*—I argue that her investment in scientific education for women has been mischaracterized as overly conservative. In fact, her articulation of a specific model of scientific pedagogy acknowledges the inevitability of existing in patriarchal structures yet encourages the measured pursuit of naturalist knowledge as a means of finding fulfillment in an otherwise restrictive personal life. Edgeworth also deploys her own extensive naturalist knowledge (earned from years of supporting her father's scientific interests and cultivating her personal passion for education) to interrogate the critical significance of botany and natural history to a woman's life—as part of, and yet distinct from, her role as an educator of young people. While natural history is primarily explored as a useful discipline for the education of children, botany, for Edgeworth and her characters, enabled an array of fruitful opportunities spanning emotional, intellectual, and social dimensions. For example, through the poignant explication of specific plants, *Belinda* positions botany as an intellectual pursuit with special significance for female participants, a dynamic reflected in Edgeworth's own botanical exchanges with her American friend Rachel Mordecai Lazarus. As a whole, this chapter exposes the enduring significance of female friendships and networks to the successful practice of naturalist labor.

The Naturalist as Artist

The second section of *Botanical Entanglements* features a different venue for women's scientific exploration: the creative visual representation of plant specimens. The most famous botanical artists from the eighteenth century are, of course, men: Georg Ehret, John Lindley, James Sowerby, and Sydenham Edwards are just a few of the most recognizable names. Whether their works of art depicted the subject matter of technical manuals or served as lavish accompaniments to full-color catalogues of flora and fungi, it was expected that the images themselves would adhere to a number of specific conventions that all added up to one rather odd truth about botanical illustration—that the final image of a given plant was intended to simultaneously look very much like that plant (or, rather, its ideal form) and very little like the actual specimen(s) used as reference for the drawing.[14] This disjuncture between utilizing one or more unique plant specimens to achieve a representation as identical to the holotype as possible and the corporeal existence of the individual organic samples is almost always elided in critical discussions around botanical artwork, both past and present.[15] The women artists of this study, however, made visible the uncomfortable erasure of the plants themselves, individual biota with personal characteristics and distinctive attributes. In this,

I locate an early female-centered anticipation of twenty-first-century critical plant studies as elucidated by philosophers and botanists.[16]

Women contributed to eighteenth-century trends in botanical representations through multiple approaches. Many worked as colorists (although often uncredited) for other artists' work in the cycle of textbook production; some with advanced artisan training even completed the sketching and engraving themselves. There was also an abundance of extralinguistic ways to explore vegetality through art. Common examples included embroidered pillows, fabric with elaborate floral motifs, and mirrows or keepsake boxes festooned with shells and/or dried flower petals. In fact, there is a long history of women's artistic engagement with natural objects throughout the eighteenth century, given its affiliation with domestically endorsed skills like drawing and painting, as well as craftwork and decorative labor. Such artistic projects have been justly celebrated by historians of material culture, but their role in the ongoing botanical discourse produced by professionals and amateurs alike merits further interrogation.[17] Across two chapters, I join scholars like Tobin in questioning the critical division between what has been understood as scientific artwork and that which has been relegated to the realm of amateur craftwork. By productively blurring such distinctions, I argue that we are better able to appreciate the full range of intellectual and creative participation in the botanical sciences undertaken by women in the eighteenth century. Further, upon allowing women's botanical artistry to speak on its own terms—instead of within the narrow confines of a technical assessment or traditional history of scientific illustration—we are able to appreciate the subtle yet radical ways in which women artists reimagined the role of the artist in relation to the individual plant subject.

In order to pursue these critical aims, *Botanical Entanglements* centralizes three women whose artistic productions represent the pluripotent possibilities in eighteenth-century botanical representation: Elizabeth Blackwell, Henrietta Maria Moriarty, and Mary Delany. Each of these women undertook botanical art projects for different reasons and under divergent circumstances, but their works reveal a common desire to find in the visual depiction of plants something greater and more profound than what was expected of floral illustration. By utilizing organic components in their compositions or making affectionate, personal reference to the model plants, for example, these women redefine the relationship of artist to subject and woman to plant. Thus, I argue that, while they adhered to the conventions of botanical artwork in many respects, their distinctive philosophical and visual approaches to their vegetal subjects must be acknowledged if we are to understand the revolutionary tactics that women used to resist the reductive narratives of female to plant dynamics.

Delany's impressive oeuvre of creative naturalist expression—including the famed mosaics—has been the subject of recent excellent scholarship, but much has been invested in celebrating the innovativeness of her technique and her relationships with prominent contemporary botanists rather than delving more deeply into the critical significance of her investment in vegetal representation.[18] On the other hand, Blackwell and Moriarty are rarely the subject of literary or cultural study in any serious fashion (and, in the case of the latter, the attention, limited as it is, is paid almost entirely to her novels).[19] While the chapters centralizing these women is in some senses a recovery project, it is also, more important, an illumination of how these (and other) women used art as a vector for interacting with and contributing to the emergent scientific field of botany. Furthermore, *Botanical Entanglements* reads these artists as cleverly navigating both scientific and artistic discourses to subvert the expected outcomes of depicting flowering plants—that is to say, where cultural scripts about women as sexualized plants (evidenced in Linnaeus's journals, made notorious by Erasmus Darwin, and deployed mockingly by a range of male authors from Jonathan Swift to Richard Polwhele) repeatedly linked female bodies with sessile, vulnerable blossoms, Blackwell, Moriarty, and Delany retold that story. Each in their own way, these artists made the argument that women could cultivate intimate, personal relationships with plants that provided fulfillment, joy, and inspiration—relationships that I argue offer new models for understanding both femininity and vegetality in the eighteenth century.

Illustrated scientific reference texts were a profitable undertaking in the late eighteenth century, given the rising popularity of naturalist study among amateur men and women alike and the avid interest in the portrayal of exotic plant and animal specimens from around the globe. From lushly colored folios like Robert Thornton's *Temple of Flora* (1799–1807) to the meticulous woodcut illustrations done by Thomas Bewick for his *History of British Birds* (1797), such illustrated works could vary significantly in cost but (unless production costs dramatically exceeded profits) could usually be relied upon to generate income. Therefore, it is unsurprising that, while such undertakings required a certain amount of upfront financial support to produce, women with an interest in botany and a talent for sketching were drawn to such projects in times of financial distress. Charlotte Smith, relying upon her sister's artistic skills, proposed such a project to her publisher in 1797, "a plan of [doing] a set of drawings, one to illustrate each of Linneas's orders—to be etched with a page of Letter press to each & the characters done with precision for the use of botanical students & those who cultivate this brand of drawing."[20]

Although Smith's own project never materialized, Blackwell and Moriarty undertook ambitious botanical projects, which star in chapter 4. Blackwell's *A Curious Herbal* (1737–39) was enthusiastically endorsed by contemporary botanists, horticulturalists, and physicians alike and served as an excellent reference for plants in the Chelsea Physic Garden with medicinal value. Moriarty's more obscure *Viridarium* (1806)—which participates in the same pedagogical discourse of scientific authority deployed by the authors of the previous chapter—emphasized the educational value in the study of flowering greenhouse plants for her audience of young female students. While their texts differ in many ways, this chapter elucidates the intimate approach each artist took to the depiction of her botanical subjects, as well as the rigorous scientific research required in order to supply the accompanying textual explications. Blackwell and Moriarty both underscored the significance of discrete biota in their paratextual material, with particular attention to describing individualized specimens (thereby defying established standards for anonymized botanical illustration). In theorizing this intimate artistic stance, I articulate a female botanical subjectivity, a model for approaching scientific study that differs from the influential concept of "modest witnessing" as championed by the Royal Society.[21] Such botanical subjectivity cultivates a deeply personal intimacy with vegetal being that is reliant upon the very corporeality of decaying leaves, bruised petals, and fecund dirt that most botanical illustration worked to elide.

Not all botanical artwork produced by women in the eighteenth century exclusively centered on painting or drawing; Mary Delany's miscellaneous collection of naturalist art projects represents the diversity of opportunities available to women for creatively engaging with plants, shells, fungi, and animals. In chapter 5, I offer a new perspective on her fruitful creation of nearly one thousand exquisitely detailed paper collages of plants, focusing on how the novelty of their form enabled her to respectfully engage with her organic subjects. By adapting the tactics of collage—a form of craftwork predicated upon the harmony of many unique constituent parts—Delany was able to recognize and respect the exuberant and excessive nature of plant life. Further, the textual characterization of her vegetal subjects reveals an enduring intimate dynamic between Delany and the individual flowers she used as inspiration for the collages, presenting yet another perspective on female-vegetal alliances. In making this claim, I contextualize these mosaics within her lifelong passion for both nature and craftwork. Studying her voluminous correspondence with family and friends over several decades, I take note of how Delany's abiding interest in scientific craftwork reveals what she understood to be the manifold possibilities for emotional, intellectual, and spiritual consolation in such activities. Her persistent emotional connection to shells, for example, reveals the ways in which conchological study for women could simultaneously fill intellectual and personal needs. Finally, a survey of her naturalist poetry provides a segue into the final section of *Botanical Entanglements*, the book's coda, as I trace Delany's abiding dedication to weaving faith and friendship into all of her naturalist activities.

The Naturalist as Poet

The coda to *Botanical Entanglements* looks to naturalist poetry of the Romantic period to illuminate one woman's literary intervention in botanical discourses. The ubiquity of nature poetry penned by William Wordsworth, Samuel Taylor Coleridge, Robert Southey, and, later, Percy Bysshe Shelley and John Keats has inspired generations of influential scholarship evaluating the divergent critical purposes to which each of these poets put their individual depictions of the natural world. Wordsworth's pensive depiction of daffodils in "I Wandered Lonely as a Cloud," for example, hints as questions of ocular perception, while Shelley's "The Sensitive-Plant" centralizes the elusive Brazilian mimosa plant, touted as evidence for plant sentience. Women's contributions to this genre have also been celebrated in recent scholarship, with particular attention paid to the ways in which women's peripheral relationship to scientific and political discourses shaped their distinctive approaches to poetic expression.[22] While Felicia Hemans and Dorothy Wordsworth both penned significant naturalist poetry, Charlotte Smith's exceptional combination of elaborate scientific detail with other forms of literary production makes her authorial career singularly important to this project. For this reason, her poem "Flora" is the subject of the book's coda.

Smith's distinctive approach to nature writing highlights the liminal position she occupied in the historical trajectory of both scientific and literary developments, with her career as novelist, children's textbook author, and poet spanning the transitional years between 1784 and 1806.

Around the turn of the century, naturalist poetry occupied a productively ambiguous space between literary and scientific discourse. While professionally authored tracts were still regarded as authoritative resources, it was common for both men and women to draw upon scientific discourses to nuance their poetic arguments. In fact, scientific information like Linnaean terms for plant parts, taxonomic data for flora or fauna, and detailed descriptions of geological phenomena could be used to bolster intellectual authority across multiple vectors.[23] With respect to botany and natural history in particular, the thriving tradition of amateur participation in scientific inquiry facilitated the easy incorporation of scientific concepts, terms, and data into creative works. Unfortunately, the increasingly orthodox approach to scientific study that arose in the early decades of the nineteenth century (accompanying a decline in the popularity of Linnaean botanical schematics) led to a disinterest in published poetry that engaged with nature on such a technical level. As a result, while poets like Hemans frequently incorporated floral imagery and scientific concepts, her work refrained from interrogating those scientific concepts further and retained the metaphorical or mythical valences of her poetic flowers as their primary attributes.[24]

In the late eighteenth century, however, the inclusion of detailed information about plants or animals served to ratify both poetic and scientific authority. John Aikin's influential tract *An Essay on the Application of Natural History to Poetry* (1777) explicated the benefits of naturalist knowledge employed in poetic discourse, delineating the growing influence of emergent sciences on the fine arts and the diverse methods used to incorporate such knowledge in verse. Naturalist knowledge could be included for avowedly educational purposes (as in the case of Frances Arabella Rowden's taxonomic guide *A Poetical Introduction to the Study of Botany*) or with the intent of dramatizing scientific ideas (as with Erasmus Darwin's two-part poem *The Botanic Garden*, which I discuss later in this book). Other poems with significant investment in the scientific underpinnings of the natural world they depict in verse—for example, James Thomson's *The Seasons* and James Grainger's *The Sugar-Cane*—deploy scientific information more subtly, or in combination with historical and cultural data. I draw attention to these two poems in particular because their use of explanatory notes has been interpreted as a means of bolstering specific interpretations of nature with scientific data. Smith's poetry and educational tracts seemingly participate in this same intellectual tradition, as the female poet underpins her botanical expertise through professional endnotes.

Botanical Entanglements departs from the critical consensus to argue that Smith's poetry, rather than affirming the value of taxonomic knowledge, actually undermines the entire concept of systematic botanical organization altogether.[25] The extensive deployment of endnotes in "Flora" obviously works to provide additional information for readers of the poem, in remarks on growth or behavior, taxonomic data, and vulgar names for plants and animals. Yet it is precisely the excess of the latter two categories with which this coda is concerned: by offering up a wealth of descriptors for a single plant, often seemingly without attention to poetic resonance or readerly familiarity, Smith reveals the unstable truth about botanical lifeforms. Where taxonomic structures labor to contain the riotous abundance of plants—endlessly expanding, infinitely variable—her subversive poetic deployment of taxonomy exposes the impossibility of such a task. Coupled with her poetic depiction of hybrid plant fairies and the goddess Flora in verse, these purposefully ineffectual endnotes reflect a critical strategy that works to forge a path forward for women and plants that transcends the patriarchal restrictions on their very being. Anticipating the kinds of posthumanist arguments that would be made almost two hundred years after her death, Smith's clever manipulation of existing scientific and poetic strictures demonstrates another facet of the distinctive philosophy toward vegetality that *Botanical Entanglements* traces throughout multiple genres of literary and artistic production in the eighteenth century. "Flora" and all of its vegetal profundity is one brief, yet significant, example of how women labored to reimagine their relationality to an organic world that stubbornly resists knowability.

In the chapters that follow, *Botanical Entanglements* works to highlight just how invested eighteenth-century women could be in scientific inquiry and what such involvement could mean for them personally and as part of a broader female-oriented community. By taking advantage of existing opportunities for creative and intellectual production—through didactic texts, art, and poetry—the women featured in this study were able to subvert the sociocultural limitations ostensibly interfering with their involvement in the broader scientific community. These authors and artists did more than simply participate in naturalist discourses, however; they meaningfully and radically reshaped the potential models women had for relating to individual biota like flowers, trees, and insects, as well as the natural world writ large.

1

Caterpillars in the Garden

Teaching Natural History in Women's Periodicals

> What Clods of Earth should we have been but for Reading?—How ignorant of every thing but the Spot we tread upon?—Books are the Channel through which all useful Arts and Sciences are conveyed:—By the help of Books we sit at Ease, and travel to the most distant Parts; behold the Customs and Manners of the different Nations in the habitable Globe, nay take a view of Heaven itself, and traverse all the Wonders of the Skies.
>
> —Eliza Haywood, the *Female Spectator*

This book begins by looking to periodicals as a venue for circulating naturalist knowledge among women readers; specifically, Eliza Haywood's *Female Spectator* (1744–46) and Charlotte Lennox's *Lady's Museum* (1760–61). As publications with avowedly pedagogical intentions in a literary marketplace that welcomed the blending of pleasure and instruction, both of these texts were able to leverage emergent opportunities for female authority in the periodical sphere in order to perform small but significant interventions in the discursive dynamic between women and nature. Such interventions begin by encouraging women to self-educate through judicious but diverse reading (a common piece of advice in a period where formal education for women ended at a very young age), but conclude by ambitiously proposing a reconceptualization of women's relationality to particular intellectual discourses—in this instance, natural history.

Although the same ideas were woven into many of her novels, Haywood's dedicated advocacy for the transformative possibilities of reading—advertised above as the ability to "travel to the most distant parts," "take a view of Heaven itself," and "traverse all the Wonders of the Skies"—is made manifest in the miscellaneous educational contents of her famous periodical, the *Female Spectator*. Along with Lennox's *Lady's Museum*, Haywood's periodical offered some of the most creative scientific content produced expressly for women in the mid-eighteenth century, surpassing in many ways the dialogic or conversational format of scientific education so popular in other critical conversations. By embedding scientific content in the avowedly edifying (and, of course, also entertaining) periodical form, Haywood and Lennox strategically engaged with prevailing ideas about women's relationship to scientific knowledge. In what follows, I assess the ways in which the *Female Spectator* and the *Lady's Museum* offer scientific content to their female readers by taking advantage of the complex relationship between women, domestic spaces, and nature.

Though similar in some respects, Lennox and Haywood adopted different modes of presenting naturalist inquiry to their readers; whereas Lennox included detailed information on discrete, specific subjects like entomology or astronomy in mini-, lesson-like segments, Haywood contextualized her endorsement of scientific information within the fictional depiction of active experimentation. In this chapter, I evaluate Haywood's critical deployment of domesticity in the *Female Spectator* as providing access to scientific subjects by circumventing traditional limitations on female scientific practice. As a point of comparison, I trace the ways in which Lennox presents a complementary strategy of studying natural history that emphasizes the enormous potential of even just a little bit of information about the natural world. Ultimately, I highlight how both authors exploit the intimate connections between woman and nature in order to encourage scientific study and cultivate interspecies sympathies.

By conceptualizing scientific study as rooted in domestic spaces, these authors endorsed a model of scientific inquiry that took advantage of the generative potential of a venue like the home garden and its manifold critical possibilities. While both Haywood and Lennox present the subject of natural history as acceptable at least initially because it does not break the boundaries of appropriate female behavior in its cohesion with domestic ideologies about female spaces and feminine comportment, each author also interrogates the expansionary possibilities of those female spaces. As I will demonstrate, Haywood uses the notion of the garden (already a potentially subversive space) to generate more intellectually fulfilling and radical practices of natural history than most male authors condoned. Likewise, Lennox configures a philosophizing walk in the garden as a metaphor for female intellectual practice and an opportunity for broader reflection on women's experience in eighteenth-century England.

Finally, I want to more fully theorize the value of locating the periodical form in a conversation about scientific innovation in the eighteenth century. An always elusive genre, periodicals were at once edifying and entertaining, simultaneously for profit and for pedagogy. It is only by grappling with these contradictions that we can critically appreciate the value of such texts to real women in the period and begin to understand the role such publications played in distributing scientific knowledge to female readers. In this chapter, I illuminate the ways in which the *Female Spectator* and the *Lady's Museum* accomplish their innovative critical aims through the careful interplay between pleasure and pedagogy.

Reading and Writing Natural History

As common as it may have been in certain intellectual circles, the proposition that women ought to self-educate through reading was beset by frequent debates regarding *what* women should read and *how* they should read. The *Female Spectator*'s seemingly self-evident claim about the value of reading was, in fact, offered in contradiction to the multiple anxieties about educated women in the eighteenth century. Reading was rare enough among women that historians estimate that only 40 percent of upper- and middle-class women in England were literate by the middle of the eighteenth century (compared to an estimated 60 percent of men).[1] As for those women who did read, the genre of their reading material was subject to frequent scrutiny and suspicion: "Botany meant sexuality, astronomy evasion of traditional femininity, classical literature usurpation of male prerogatives, poetry a disruptive imagination, metaphysics revolution, the novel seduction."[2] While such anxieties represented a more traditional viewpoint on the subject, they were still an active presence in a culture that relied upon the understanding of women as moral creatures (while simultaneously decrying them as *immoral* ones; Alexander Pope's declaration that "ev'ry Woman is at heart a Rake" is only the most famous of such condemnations).[3] As Jacqueline Pearson laments, "It is almost literally true that all genres of reading, however apparently safe, upset some commentator in this period."[4]

However, in spite of a few vocal detractors, by the early years of the eighteenth century, new ideas about education for women incorporated naturalist topics. While the *Athenian Mercury* (1690–97), the *Ladies' Diary* (1704–1840), and the *Free Thinker* (1718–21) all presented scientific topics to women in some significant respects, the dissemination of such material was almost always limited to a basic question-and-answer interaction, in which real or fictitious elementary questions on scientific subjects submitted by women would be publicly addressed by eidolons of the periodical (who were often male, although sometimes female). In contrast, periodicals written by female authors for an expressly female audience had the potential to offer more daring models of women's scientific interests. While a few other female-authored periodicals in the mid-eighteenth century were invested in the subject of education (sharing many qualities with the conduct manual genre), Haywood's *Female Spectator* and Lennox's *Lady's Museum* were two publications that focused a significant part of one or more issues on scientific subjects in the context of that discussion.

Kristen M. Girten cites the long-running publication the *Ladies' Diary* to note the prevalence of scientific content in periodical literature in the mid eighteenth-century, explaining that much of the material "evokes what had become a common assumption by the time of its publication—that natural philosophy is a valuable pursuit of the fair sex."[5] It was, moreover, clearly a pursuit that interested a significant portion of the educated English population, likely because it was viewed as more approachable to a lay audience; the sheer number of specialized encyclopedias and dictionaries printed during this period attests to its material value in the eighteenth century.[6] Natural philosophy was a field with the potential to be tremendously diverse, but it was also often subject to a gendered censorship, in which female audiences were allowed access to only the most basic of its tenets and practices. As I will elaborate upon in later chapters, botany in particular was an especially problematic field for women because of its metaphorical and imaginative connections to human sexuality.

Still, the study of insects, animals, and plants—usually grouped under the umbrella of natural history—was commonly encouraged as a harmless and appropriate pursuit for women, one that would not take a woman of the leisured class out of her social sphere. The level of education or exposure to scientific material required to study a caterpillar under a microscope was minimal, and most popularizing texts provided that information in easily digestible, simplified terms. (Benjamin Martin's publications were among the most frequently reprinted of this genre.) Such study was popular with diverse groups of women, including Queen Caroline and Augusta, Princess of Wales, who played a part in the founding of the Royal Botanic Gardens at Kew, as well as women of the merchant class who were members of the botanical club that met at the Temple Coffee House.[7] Provided that the study of natural history did not distract a woman from her otherwise required duties, it was a far better pastime than gossip, walking in public parks like Ranelagh, or shopping. It is unsurprising that the *Female Spectator* and the *Lady's Museum* would engage with the subject, given its popularity and generally positive reputation.

Like many of the mid-century educational subjects marketed to women, natural history was also understood as conveying a moral benefit. George Adams, in the popular *Micrographia Illustrata* (1746), loosely explains the subject as "the Study of Nature, or in other Words, a serious Contemplation of the Works of God."[8] To the eighteenth-century mind, the study of nature was inevitably bound up with notions of religion, as Haywood's and Lennox's writings demonstrate. The literary linkages made between nature and religion (also known as natural theology) reflected a "conventional scientific piety that helped facilitate women's access to natural history."[9] This connection would be more overtly realized with texts like Sarah Trimmer's well-known *Easy Introduction to the Knowledge of Nature, and Reading the Holy Scriptures, Adapted to the Capacities of Children* (1780), in which the sole value of naturalist study is its connection to religion. In some respects, this deliberately perpetuated connection between religion and natural history stemmed from the period's obsessive concern with policing not only the physical but also the mental behaviors of women. As women's education and notions of appropriate feminine deportment became emblematic of and essential to the eighteenth-century construction of a new bourgeois identity, anyone advocating for female education had to be very careful to avoid the appearance of intellectual transgression. However, as readings of the *Female Spectator* and the *Lady's Museum* will demonstrate, periodicals could offer scientific knowledge in a way that acknowledged the social limitations to women's intellectual practice while offering alternative modes of studying natural history that subtly defied those limits.

At the same time, scientific content in periodicals was (and often still is) considered ancillary to its primary aims and less rigorous than scientific content in other venues. While it is seemingly true that the valuation of professional or academic scientific texts as superior to consumer-oriented scientific texts was as pervasive in the eighteenth century as it is today, it is also the case that critics have the opportunity to rethink this dynamic in assessing the value of public-facing texts to an underrepresented population. Barbara Gates and Ann Shteir (drawing upon Roger Cooter and Stephen Pumfrey) suggest that critics "treat popular science as its own form of knowledge, shaped in relation to the needs of audiences beyond elite and learned culture": a proposition that affords us the flexibility to more accurately understand the function that popularizing texts served for women readers and writers of the period.[10] To write a botany textbook for children or series of essays on entomology in a financially profitable journal should not be understood as amateur work, but, rather, as distinctive means by which women authors participated in scientific discourse. Further, to read and engage with such a production should not be interpreted as passive consumption, but as a response to an active call to experimentation that allows women readers to make the most of their reading experience and to use that experience to cope with what might be an otherwise intellectually restricted personal life. Reading these periodicals by women as not merely imitators or distributors of scientific knowledge but as crucial participants in eighteenth-century scientific discourse ultimately expands the archive of women authors writing about science and reevaluates the role that women's periodicals played in scientific education.

Women and Nature in Eliza Haywood's *Female Spectator*

The *Female Spectator* is among the most significant periodical publications of the eighteenth century, not least for its innovative discussions of a wide range of issues pertaining to women's experiences and education. Across the two years of its publication history, Eliza Haywood wrote immensely popular issues that addressed everything from romance and courtship to politics, religion, and science. One of the most remarkable of these topics is her active endorsement of female scientific practice, initially articulated through the male persona Philo-Naturae. In his dialogue, she presents a perfunctorily conservative position that the female eidolon converts to a subtly rebellious undertaking, ultimately disrupting conventional narratives about women's relationship to nature. Through her inventive portrayal of women engaging in naturalist work, Haywood cannily interrogates the relationship between women, their scientific subjects, and the spaces in which scientific labor is undertaken. A careful study of this dynamic allows us to understand the distinctive way in which Haywood conceptualized the affiliation between women and nature, a critical reconfiguration that likely influenced Lennox, Charlotte Smith, and others later in the century.

Like other essay periodicals, the *Female Spectator* used the fiction of an editorial board, in her case comprised of female characters who represented various ideal models of femininity: the experienced but respectable spinster the Female Spectator, who functioned as the periodical's primary voice; the witty, young, devoted wife Mira; an unnamed "Widow of Quality"; and the charming and accomplished Euphrosine.[11] The *Female Spectator* was a monthly publication, issued in twenty-four parts, between April 1744 and May 1746, and was popular enough to be sold as a bound volume of multiple issues as early as 1748, while single issues of the periodical were still being distributed.[12] In fact, Patrick Spedding has concluded that the *Female Spectator* was Haywood's single best-selling publication of her career, earning her about sixty guineas per year.[13] Moreover, as Lynn Marie Wright and Donald J. Newman note, "the fact that [the *Female Spectator*] lasted two years when many other periodicals lasted only a few weeks or months argues that Haywood had a wide and loyal readership. We know she had audiences in Germany, France, Italy, and as far away as America."[14] Additionally, a number of Haywood's later periodical publications, including the *Parrot* (1746) and *Epistles to the Ladies* (1749–50) were written "by the authors of *The Female Spectator*," indicating a publisher's interest in capitalizing on her success. Several of Haywood's later publications even used characters from the *Female Spectator*: Mira was the narrative persona in the *Wife* (1746), and Euphrosine was the same in the *Young Lady* (1756). Haywood also recycled these characters for use in *Epistles for the Ladies*.

While critics have often been divided on the subject of Haywood's intended audience, I agree with the assessment that whatever her actual audience demographics may have been, her intended readership was largely female.[15] This is not to say that Haywood expected an exclusively female audience; on the contrary, a large number of the correspondents were male (including Philo-Naturae, the primary scientific interlocutor of the periodical), and much of the Female Spectator's advice on courtship and marriage was directed at both sexes. However, a repeated tactic throughout the periodical relies upon the flexibility of a female intellectual voice that speaks without seeming to disrupt male intellectual authority in order to relay a substantive critique (often of that male authority). Haywood writes, "If therefore I have directed my Advice in a peculiar Manner to those of my own Sex, it proceeded from two Reasons, First, because, as I am a Woman, I am more interested in their Happiness . . . secondly, I had not a sufficient Idea of my own Capacity, to imagine, that any Thing offered by a *Female Censor* would have so much Weight with the Men as is requisite to make that Change in their Conduct and Oeconomy, which, I cannot help but acknowledge, a great many of them stand in very great need of."[16] While certainly tongue-in-cheek, Haywood's qualification in this passage can be taken to represent the consistent limitations placed on women's intellectual capacities that the Female Spectator must circumvent. And, by making a gendered appeal to her audience, she rhetorically positions herself and the eidolon as less ambitious or assertive than other passages might suggest. Deflecting the kind of sexist criticism that seemed to plague Haywood (among many other women in the eighteenth century), this passage also offers a subtle critique of those who would ridicule or shame a woman for assuming intellectual authority.

Periodicals were among the forefront of those publications advocating for and enabling women's scientific study, beginning with the aforementioned publication the *Lady's Diary*, which presented scientific and mathematic content to female readers. A few of the more popular naturalist texts in English during this time could have been and were likely read by some women, including: John Woodward's ambitious *An Essay toward a Natural History of the Earth and Terrestrial Bodies* (1695), Patrick Blair's highly technical but very useful *Botanick Essays* (1720), Philip Miller's frequently reprinted *The Gardener's Dictionary* (1731), and Joseph Miller's *Botanicum Officinale* (1722), which was published in English despite its title. Texts like these were available to female readers of varying financial means and different social classes and were occasionally even explicitly directed toward female audiences. Women were also renowned as translators of popular Continental scientific texts designed for female readers, including Aphra Behn (who translated Bernard le Bovier de Fontenelle's *Entretiens sur la pluralité des mondes* in 1688) and Elizabeth Carter (whose translation of Francesco Algarotti's *Il Newtonianismo per le dame* in 1739 won her high praise). That Haywood would choose to address scientific study in her multifaceted periodical was on the surface unremarkable; what made her elucidation of women's naturalist practice so inventive was its acknowledgement of the significant relationship between women, nature, and domesticity. As I will expand upon below, women were typically understood to be intimately connected to domestic spaces and the natural world through various associations between femininity, the home, and nature (specifically, flowering plants). These affiliations were often used as justification for limiting women's involvement in public life and meaningful intellectual labor. When turned to subversive purposes, however, these connections had impressive potential.

Pamela H. Smith, among others, has traced the extent to which science in the seventeenth and early eighteenth centuries was influenced and shaped by domestic spaces and practices—especially domestic craft and the production of household goods.[17] Early natural philosophers such as Robert Boyle were known for their home laboratories, and male amateur practitioners across multiple scientific fields used domestic spaces such as kitchens, distilleries, and gardens for their experiments. Many materials and ingredients required for certain experiments, for example, were easily found around the home for culinary or medicinal purposes. For women, however, recourse to domestic space for intellectual and naturalist endeavors was inextricable from their own rootedness in domestic ideologies: women were linked to the home in certain intractable ways both literal and cultural, and their participation in public spheres of science suffered for it. Fortunately, their amateur practice of natural history was enabled by domestic spaces just like the ones evoked in the *Female Spectator,* as women had the potential to collect specimens, make and record observations, and draw conclusions without the need for expensive technology or professional spaces.[18] Likewise, the cultural connections between women and craftwork of the kind Smith explores—as well as projects that utilized natural specimens such as shells and plants, studied in chapter 5—strengthened the case women could make for participation in naturalist study. In the *Female Spectator,* the eidolon and her companions leverage the association of women with domestic spaces (gardens and parlors) and domestic subjects (snails and caterpillars) to pursue innovative naturalist projects. Through such scenarios, Haywood's articulation of the practice of natural history keenly elucidates the connections between domesticity and its facilitation of female scientific practice in the eighteenth century.

Scientific labor is a key component of book 15, the issue of the *Female Spectator* in which Haywood articulates her vision of the connections between women, nature, and natural history. The eidolon expends more textual capital on natural history than any of the other subjects of study she encourages, and, along with astronomy, it is the only one that she and her fellow committee members actively pursue. She introduces the subject in the course of a conversation on taste, presenting the study of natural history as an alternative to a woman's typical fascination with dress and gossip: "How much more Pleasure should we find in examining the various and beautiful Habits with which Nature cloathes those Plants and Flowers which adorn our Gardens," in comparison to "the Dress of every one we see at the Ball, the Court, the Opera."[19] This subject is initially introduced within the context of her avowed reformatory project, thereby framing it as a solution to female vanity and idleness, as well as a visual alternative to material culture. The "Plants and Flowers" of a woman's home garden are more appropriate subjects for her contemplation than the French lace or jewels of her companions, evoking also the value of natural over artificial beauty, a subject of great concern in eighteenth-century aesthetics. At first glance, natural history is presented in a restricted way to women: offering as objects of study only the simple, harmless subjects of the garden that might be encountered on an afternoon walk.

In the initial letter by Philo-Naturae, Haywood expands upon the topic in more detail, as the male correspondent enthusiastically but narrowly recommends the subject to the readers and editors of the *Female Spectator.* He introduces natural history as a means of enjoying "the Odours of ten thousand different Flowers," but he hastily circumscribes the subject as one of experience rather than devoted study:

> Contemplation on any thing may be called a Study in more or less Degree, I would not be thought to recommend to the Ladies . . . that severe and abstruse Part which would rob them of any Portion of their Gaiety:—On the contrary, I would not advise them to fill their Heads with the Propositions of an *Aldrovandus,* a *Malbranche,* or a *Newton:*—they require a Depth of Learning, a Strength of Judgment, and a Length of Time, to be ranged and digested so as to render them either pleasing or beneficial.[20]

This passage encapsulates a number of the social limitations placed on women's intellectual activity in the mid-eighteenth century: the belief that intense study was unfeminine; the equally persistent belief that it was beyond the intellectual capacities of women to pursue it at a profound level; and the common distillation of current scientific principles into a form deemed accessible for women readers. As Girten points out, "His advice appears to comply with, if not enforce, the convention of confining women to the private domain" while still advocating for some small progress, a dynamic within these sections of the *Female Spectator* that Girten explicates as an opportunity for women to "expand the horizon of, and thereby reshape, the sphere to which they are consigned."[21]

However, Haywood's choice to locate the ladies' scientific efforts in the domestic sphere accomplishes more than simply reconfiguring their experience of domesticity. Their claim to a specific kind of empirical authority, while absolutely entertaining and enlightening (as Girten argues), is also a challenge to the perceived association between women and nature, in which women, like natural objects and spaces, are subject to the authority and expertise of men. Haywood's periodical usefully complicates this presumption, reflecting the more recent observations of historians of science that such dramatic disparity between masculine and feminine scientific practice was more wishful thinking on behalf of certain philosophers and early historians than reality.

To return to Philo-Naturae: it is possible to read his rejection of the rationalist philosopher Nicolas Malebranche or the polymath Isaac Newton as Haywood accepting some limitations to women's scientific study, but the inclusion of botanist and natural historian Ulisse Aldrovandi complicates such a supposition. As a man who was widely regarded as one of the most influential naturalists of the early modern period, Aldrovandi would have been a promising figure for women to learn about even in a casual pursuit of natural history (albeit one whose treatises were written exclusively in Latin). The founder of one of Europe's first public botanical gardens (in Bologna in 1568), he was also an important early contributor to the widespread contemporary interest in natural history as a publicly accessible science.[22] As the women of the periodical engage in naturalist projects, they do so in a cultural moment when widespread interest in natural history was inspired by the legacy of such sixteenth-century research: the invocation of Aldrovandi thus implicitly endorses their naturalist labor by tying the female practice of natural history to a scientific tradition. By disrupting the correspondent's cautionary remarks through such inconsistencies, Haywood deftly creates space to navigate the seemingly unavoidable tensions between women's actual intellectual capacities and what men thought those capacities could or should be. Fortunately, this initially restrictive vision for natural study is upended when the eidolon pursues active experimentation. The Female Spectator declares, "Our little Society, who have agreed to pass a few Days at a Country Seat, belonging to our President, the excellent *Mira* . . . to make those Inspections. . . . It must certainly be confessed, that there is nothing more entertaining, or more profitable to the Mind, than the Study of Natural Philosophy."[23]

Active experimentation, for the Female Spectator, is crucially linked to specific organic spaces—spaces that convey important information about the ladies' naturalist endeavors. While there is virtually no description in this letter of Mira's estate other than its vague location in "the Country," the Female Spectator's hints about their rural getaway include "Apple Trees . . . Cabbages, and several Plants in the Kitchen-Garden" and a "Parterre," which suggests that there are actually multiple different gardens through which the ladies range in pursuit of their naturalist speculations.[24] Encompassing both a pragmatic agricultural site like a kitchen garden and a highly stylized, purely ornamental garden component like a parterre, Mira's estate elicits a range of botanical and cultural associations. The parterre is most typically an ornate, raised garden bed that could be filled with any number of cultivated exotics, but it was also going out of style by the mid-eighteenth century in favor of naturalistic English landscape gardens, as exemplified by the work of Lancelot "Capability" Brown.[25] Haywood's aside that "the Ground had lately been thrown up in order to make some Alteration to one of the Parterres," while potentially indicating a transition away from this older style, also, and more important, reflects the interruption of idle aesthetics with active scientific pursuit. That is, the fact that a dug-up parterre is where the ladies discover a collection of immature snails that later provide the basis for some of their major naturalist observations reflects the hierarchy of intellectual values that Haywood outlines in this text. Furthermore, the focus of the ladies' attention is squarely on the "little living Creatures" writhing around in the dirt, and not on the new and exciting design changes in Mira's garden.[26] While attention to the visually appealing "Plants and Flowers which adorn our Gardens" has already been endorsed by the Female Spectator, this preference for grubby little insects over showy displays of wealth reinforces the women's scientific aspirations.

Haywood's choice of the garden as the locus for the ladies' scientific investigation also reflects a particular attention to the function and significance of the garden in scientific thought. The garden was a common site of learning for naturalists, as both men and women examining nature might conduct their observations in gardens—including the illustrator and author Elizabeth Blackwell (a subject of chapter 4), whose magnificent *A Curious Herbal* (1737–39) was produced with intimate access to the elite Chelsea Physic Garden. In evoking the garden, Haywood also deploys a common trope in scientific literature of the period by alluding to texts like Behn's aforementioned translation of *Conversations on the Plurality of Worlds*, in which the curious marquise is educated in scientific subjects by a male philosopher as they stroll around her opulent gardens. Women looking to male authorities for a source of scientific information within the garden space reflected a conventionally gendered hierarchy of knowledge in which the man is the scientific authority, passing on his wisdom and knowledge to the woman, while simultaneously putting the woman in a position of relative power as the proprietor of that garden. A male correspondent may have been the one to propose naturalist study to the Female Spectator, but it is Mira whose garden they ultimately choose as their site of inquiry.

Philo-Naturae's proposal, however, is more than a fleeting recommendation. Haywood expands her portrayal of female scientific labor as the correspondent urges not merely the casual inspection of flowers in the garden, but the intentional pursuit of their more scientific details. He offers the following proposition: "As Ladies frequently walk out into the Country in little Troops, if every one of them would take with her a Magnifying Glass, what a pretty Emulation there would be among them, to make fresh Discoveries?—They would doubtless perceive Animals which are not to be found in the most accurate Volumes of Natural Philosophy; and the *Royal Society* might be indebted to every fair *Columbus* for a new World of Beings to employ their Speculations."[27] In this paradigm, women can employ their time in the study of natural history as part of an already acceptable feminine exploration (a social outing to the country) and in ways that offer potentially radical results—the discovery of something even the Royal Society might have overlooked. This brief mention of the chance to compete with "Volumes of Natural Philosophy" is a surprising moment in Philo-Naturae's narrative, and one that begs further assessment. Given that he has already rejected "study" as unfeminine and the reading of texts by members of the Royal Society (like Newton) as too difficult, why does he intimate that a woman in the casual pursuit of natural history has the potential to become a "fair *Columbus*"?

Further attention to the periodical reveals Haywood's clever subversion of her correspondent's limitations. While Philo-Naturae evokes the fantasy of a "fair Columbus" at least in part to underscore the imitative nature of the women's explorations, the female eidolon's active practice of scientific investigation embraces the true spirit of the epithet. Haywood seeks in this passage to reappropriate the fantasy of Columbus as a pioneer in exotic explorations, instead deploying the Italian as a source of inspiration for women whose journeys only take them as far as the limits of an English garden. Haywood's use of the epithet, while uttered potentially ironically by the male correspondent (or in tension with his other statements, as Girten proposes), is taken seriously by the female naturalists.[28] The country garden is no longer an inert ecological backdrop to the idle curiosities of women, but a representation of the frontier in itself. While the geographic horizons of many male naturalists were broader than those of most women, Haywood reimagines the garden as itself a potentially endless space of inquiry that is at least ostensibly outside the purview of male naturalists.

Such potential for boundless discovery is also coupled in this text with Haywood's endorsement of female empirical abilities. The social, cultural, and economic coupling of women and domesticity usually afforded women some measure of authority over the domestic sphere. However, while some interpretations of the *Female Spectator* (including work by Katherine Rogers, Kathryn Shevelow, Patricia Meyer Spacks, Shawn Lisa Maurer, and Ricardo Miguel-Alfonso) argue for the author's preservation of a conservative vision for female deportment, as exemplified by this emphasis on domesticity, scenarios like those elucidated in book 15 require us to nuance such an interpretation.[29] As Sarah Prescott and Jane Spencer argue, Haywood's articulation of feminine authority in the persona of the Female Spectator "was wielded over a wide range of public issues" and in fact did not "[dwindle] into domesticity" in spite of these reductive claims.[30] Likewise, Eve Tavor Bannet has compellingly presented Haywood's "promotion of women's learning and love of reading" as far from "[subordinate] to masculine control," but, rather, capable of producing and writing about subjects within which men are given "a purely instrumental and subservient role in relation to the female world."[31] Building upon Bannet's assertion, I propose that the conversation around natural history is a space in which Haywood not only provides valuable knowledge to her female readers but also articulates a distinct vision of the role of female empiricism. I have already mentioned Girten's argument that empiricism in this text has the power to "liberate [women] from the confines of their quotidian existence"; what I propose, further, is that female empiricism can compete with and counter conventional scientific knowledge.[32] By depicting a naturalist expedition undertaken within the explicitly domestic space of Mira's kitchen garden, Haywood gives Mira and her companions an edge when it comes to the knowledge gained therein. Women as individuals with special authority over the domestic here also have special authority over this scientific space—as will become more evident during their observations of immature snails. Philo-Naturae can encourage their pursuit of naturalist study in the garden, but he cannot control their experiments; their unnamed gentleman companion can propose a theory of insect coloration, but, as I will explore below, his incorrect theory is appropriately rejected.

Female empirical abilities are reinforced by the ladies' willingness and ability to use optical technology. The Female Spectator grandly declares in favor of the truly spectacular powers of microscopy: "All those Curiosities, which are discoverable by the naked Eye, are infinitely short of those beyond it:—*Nature* has not given to our Sight the Power of discerning the Wonders of the minute Creation:—*Art,* therefore, must supply that Deficiency:—There are *Microscopes* which will shew us such magnificent Apparel, and such delicate Trimming about the smallest Insects, as would disgrace the Splendor of a Birth-day."[33] Unlike Margaret Cavendish's famous diatribe against optical technology in *The Blazing World* (1666)—in which both telescopes and microscopes are condemned as "false informers" that do little more than "delude the senses"—Haywood describes the relationship between women and microscopes as a positive one.[34] Because the organic capacities of people are insufficient to "[discern] the wonders of the minute Creation," a refusal to take advantage of such technology would prevent the appreciation of "such magnificent apparel, and such delicate Trimming . . . as would disgrace the Splendor of a Birth-day." Moreover, the invocation of women's "Birth-day" apparel also generates for the readers a compelling association between the bodies of women and the bodies of insects—a connection that will become more significant later in the same issue.

In spite of such grand declarations about the potential for new discoveries and exhilarating interactions with nature, the female practice of natural history seems to falter as the Female Spectator recounts the story of their weekend at Mira's country estate. For one, the breathtaking visuals and vast potential for exploration extolled by the Female Spectator and Philo-Naturae are nowhere to be found: "The Weather . . . as altogether unpropitious to our main View in going into the Country . . . the unusual Cold, and almost continual Rain, or what was even worse, a kind of poisonous Dew that sometimes fell, had either destroyed great Part of those little Creatures . . . or else had driven them to Shelter."[35] This marks an odd hesitancy to portray the fullest potential of the intellectual practices that Haywood recommends. Girten makes the convincing suggestion that Haywood uses these illustrations of interrupted practice to parallel the general confinements of proscriptive femininity; I would expand upon that comparison to argue that Haywood creates fictional moments of opposition to proscriptive femininity in order to illustrate a woman's ability to overcome and surpass these limitations.[36]

The practice of natural history—rain or shine—becomes an instrument of rebellion not only against the patriarchal system that restricts female intellectual practice but also in opposition to the male-authored vein of scientific literature that persists in diluting scientific knowledge in order to limit and ultimately control women's intellectual labor. Thus, Haywood's protagonists venture out in spite of the rain; the women remain undeterred and unearth a few caterpillars and snails, commonplace creatures that provide a vital opportunity for the Female Spectator to display a detailed knowledge of their physical appearance and behavior. Far gone is Philo-Naturae's pleasing fantasy of young ladies traipsing through fields of flowers, now replaced with a determined group of women trudging through mud; the practice of natural history transcends its connotation as an activity to enhance an idle walk in the garden and becomes instead an activity to be pursued for its own sake.

The subject of the ladies' study is important, too. The correlation between women and caterpillars is compelling: both are ostensibly insignificant in the general assessment of things but important in the sympathetic worldview the Female Spectator has been working toward. Philo-Naturae primarily makes the connection between women and caterpillars explicit in a derogatory way; he jokes, "Even those Worms which appear the most despicable in our Eyes, if examin'd into, will excite our Admiration . . . when one Form grows withered and decayed, the happy Insect has another in Reserve, and, shaking off the old, appears again in all the Freshness and Vigor of Youth:—What would a certain Lady . . . and many other antiquated Beauties, give, they had the same Power?"[37] The Female Spectator, while having certainly critiqued women elsewhere in the periodical for their inappropriate vanities (including a critical reflection on her own youthful vanities in book 1), is more interested in the potential joy and enlightenment that the study of insects offers to curious women. Caterpillars, observed in "great Numbers," are readily presented as potential subjects for the Female Spectator and her companions. She muses, "As there are a vast Variety of these Creatures, I think the Learned say, no less than upward of three hundred different Species, which yet all pass under the same Name, one cannot help admiring the Wisdom and Justice of Nature, which has bestowed her Bounties" upon such insignificant little creatures.[38] In this extract, Haywood's text reflects the conventional piety of natural theology, a belief that Philo-Naturae makes explicit in his introductory missive. While this theological bent is certainly important in the text, for Haywood the study of nature is also about profound interactions between the naturalist and her subjects.

Indeed, she argues assertively for the value of studying the diminutive in nature, defensively framing the following accusation as reflective of an inability to appreciate the wonder and significance of small things: "I hear some People say, *Could they find no Objects more worthy their Attention, then Caterpillars and Snails?—Two Insects the most contemptible of any.*"[39] These "contemptible" insects, however, provide ample opportunity for the women to exercise their empirical abilities and demonstrate their scientific acumen.[40]

Those empirical abilities are highlighted most significantly through the careful observations that the Female Spectator and her companions make about the caterpillars in Mira's various gardens, including those of a "dirty brown Colour" with a "shaggy Coat" and "Bristles [like] a Boar."[41] Her notes on this caterpillar specimen echo the kind of precise, succinct empirical detail typical of naturalists including Maria Sibylla Merian, Caspar Bauhin, and Linnaeus.[42] Indeed, her observations, like those of her predecessors, go beyond the purely visual, as she notes matter-of-factly: "[They] have very sharp Teeth I experienced, by laying one of them upon the Back of my Hand . . . they have a great Number of Feet . . . [and] I perceive the chief strength of these is in those two that are placed at the Extremity of the Body, and have so much Elasticity in them, as to enable the Creature to raise itself almost upright."[43] Here, the caterpillar is not merely an object transiently noticed on a stroll along garden paths; it is the center of her scientific attention, a species whose exciting and novel abilities have inspired avid intellectual curiosity in its viewers and in the periodical's readers. The caterpillar is important to the Female Spectator as a scientific subject and as an object of curiosity; it is important to Haywood for the opportunity it gives her (as the author) to portray female empirical abilities as occasionally superior to those of their male interlocutors.

The ladies' early forays in the country, in fact, establish a marked difference between their approach to naturalist study and that of their occasional male companion. Haywood writes, "[A Gentleman] told us, that this Species of the Caterpillars is of the *Camelion* kind, and changes its Hue according to the Weather:—If we had continued in the Country a little longer, I certainly would have made the Experiment, by keeping one of them in a Box, with some Earth, and the same Sort of Leaves on which I found them feeding . . . I am rather apt to believe the Colour of these Animals more owing to their Food than to the Air they breathe."[44] This anonymous gentleman is introduced to the readers as "the worthy Gentleman . . . who is a great Contemplator of the minute Works of Nature," implying that he is also an amateur naturalist. However, the Female Spectator, who has already been established as a city dweller with little experience in the natural sciences, is readily able to provide an alternative explanation for the hue of their caterpillar subjects. Unsurprisingly for Haywood fans, while the man is positioned as the voice of scientific authority on their expedition, the woman's hypothesis most closely approaches contemporary understandings of cryptic coloration in insects. The understated aside—"I am rather apt to believe [my own theory]"—only reaffirms the value of female empiricism: the Female Spectator's outlined experiment would have been the most intellectually sound method for solving their caterpillar puzzle, and the fact that it is a woman rather than a man who proposes such experimentation is key.

The Female Spectator and company continue to display their empirical acumen upon capturing a few snails from the disrupted parterre. Although the experiment is a simple one, it combines empirical principles of natural history with the domestic setting in a way that was unanticipated by Philo-Naturae's remarks. The snails that they discover are immature, "encased in Shells, which seemed exactly the same with those of Snail, though of a different Colour and almost transparent," so the ladies decide to monitor them for a few days: "To be assured, if possible, what they were, we put a sufficient Quantity of Earth into a Pot, and then laid them lightly into it, strewing a few Fine Leaves on the Top, and carried them into the Parlour, with a strict Charge to all the Servants not to remove it from its Place, nor suffer any thing to fall upon it, or crush the Earth."[45] These women have uncovered Philo-Naturae's "Animals," brought them into the home, and proceeded to set a detailed watch on their development. Over the remainder of their stay in the country, they continue to reexamine the snails each day, taking note of changes in color, width, and the development of their tentacles.[46]

The placement of this habitat in a parlor—an explicitly female space within the already feminized home—while underscoring their amateurism must also evoke the connection Haywood repeatedly makes between domestic space and a woman's practice of natural history, as well as the connection between women and small animals. As tiny creatures readily identifiable through their affiliation with the concept of home, snails are also appropriate subjects of study for women. Like women, snails are considered biologically less important than other creatures while simultaneously functioning as both moral exemplars *and* cautionary tales: snails embody the virtue of patience or the sin of sloth, while women are either virtue made flesh or the incarnation of sin. It also seems that snails were linked to public expressions of feminine intellect at midcentury. A short-lived periodical called the *Snail, or the Lady's Lucubrations* (1745) by the pseudonymous "Eloisa" avowed its dedication to honesty, virtue, and wisdom (while retaining the governing premise of being "chiefly design'd to be an Amusement to the Fair Sex").[47] Within the pages of the *Female Spectator,* snails inspire a vital interspecies affinity, as the women care for these immature snails the way they might care for infant humans: the eidolon muses, "I must confess I am a little interested in these Animals . . . [I] had the breeding up of some of them."[48] She even genders their snail specimens as feminine, noting keenly, "The Shell . . . serves her as a House or Cavern, in which she may either hide herself, or peep out of, as she pleases, as Occasion requires."[49] The snail is not only explicitly demarcated with feminine pronouns; she is also assigned archetypal feminine characteristics such as shyness. Haywood thereby reaffirms the connections between the adventurous ladies and their delicate experimental subjects. Snails, however, also served as key components in the constitution of cosmetics like snail water (and are still found in twenty-first-century serums, creams, and gels that incorporate snail mucin extract).[50] As a result, many women in the eighteenth century relied upon the bodies of snails to enhance their own bodies, a dynamic that must invariably color this vignette of women dutifully babysitting a few gastropods.

Although presented in substantially less detail due to the nature of her periodical format, the Female Spectator's naturalist experiments echo in basic form comparable experiments performed by professional naturalists. For example, the eidolon provides the following details on their snail subjects: "We then took one [of the snails] out, and found it considerably increased in Bulk, and that the Shell was grown harder, and of a more brown Colour, and could now discern those four Antennas, or Horns, as they are vulgarly called, but are in reality jointed Tubes, which they can either extend or contract at Pleasure."[51] In comparison, I look to the December 1740 issue of *Philosophical Transactions,* in which the Dutch naturalist Job Baster recounts an observation of some destructive worms in Holland: "The Head is of a most wonderful Structure, being covered with two hard . . . (I know not which to call them, Shells or Hemicrania) of a Substance neither testaceous nor osseous, securing their softer Contents."[52] While Baster goes on to dissect his subjects and illustrate their bodies in great detail for his readers, the Female Spectator has given her readers an equally illuminating description of whole bodies in her text—another sympathetic connection between women and their diminutive insect subjects. By elaborating upon not only the subject at hand but also the experiment undertaken to explore that subject, the *Female Spectator* shares certain relevant characteristics with *Philosophical Transactions*. In fact, where Philo-Naturae has rejected female participation in the Royal Society's intellectual project—advising against reading Newton, for one—the Female Spectator has enthusiastically embarked on a series of observations and experiments that echo the kind of naturalist undertakings of Royal Society members. The "fair Columbuses" have largely ignored Philo-Naturae's conservative cautions and have instead pursued an empirical course of study that rivals, as he suggests slyly, the results produced by naturalists like Baster.

In fact, the study of "contemptible" insects was popular with a wide range of Royal Society members and correspondents in this time period, including John Bartram, who studied mussels in Pennsylvania; Abraham Trembley, who worked extensively with freshwater polyps; and the latter's nephew Charles Bonnet, who published an excerpt of his ongoing observations on terrestrial insects in *Philosophical Transactions* in December 1743. The excerpt opens with observations on caterpillars, in which Bonnet marvels at their biological ingenuity. He writes, "Nothing surprises more in Insects than their Industry; and *Caterpillars* yield to none in this respect: Not to speak of those which build for themselves Sheaths or Cases, in which Silk, their own Down, Bits of Bark, Pieces of Paper, &c. are so artfully wrought together."[53] The Female Spectator, likewise, makes a series of observations about the impressive biology of caterpillars, noting particularly, "We found [some caterpillars that] had a Sort of Glue within their Bowels, by the Ejection of which they could, when any Danger . . . threatened, fasten themselves so firmly to the Bark of a Tree."[54] Both extracts exemplify the attention to detail and passionate approach to observing insect specimens that was expected of naturalists in the period and can hardly be distinguished from one another once taken out of context. Haywood's "fair Columbuses" are fulfilling all the potential of that epithet in ways largely unanticipated by Philo-Naturae.

With the publication of the *Female Spectator*, Haywood achieves one of the most important goals in naturalist study of the period: the circulation of her scientific ideas and observations. As Mary Terrall's study of early eighteenth-century entomological labors reminds us, "The production of [naturalist] knowledge relied on material and intellectual input from different places and different kinds of people."[55] Far from being a popularizing text that made no original contributions to the field of natural history, Haywood's intellectual input (or any input from female naturalists) "could only benefit the rapidly expanding science of natural history," as Terrall points out.[56] Likewise, the labors of the Female Spectator and her companions are hardly confined to the limited sphere of action that Philo-Naturae proposes; as my analysis of their naturalist experiments and observations has shown, their canny utilization of domestic spaces as a means of generating a distinctive kind of epistemological authority reaffirms the significance of even the most basic observations of those "contemptible" creatures to the naturalist community.

It would be inaccurate to argue that the *Female Spectator* participated in scientific discourses of the period in the same way as *Philosophical Transactions*. Although Haywood's periodical was very successful in the print marketplace, it could not compete with this professional scientific publication with respect to scope of subject matter, range in readership, complexity of content, or influence within the broader scientific community. Disciplinarily speaking, it might also be fair to say that the periodical's use of the category "natural philosophy," a concept most ardently championed by the Royal Society, was aspirational. This term typically encompassed a broad spectrum of scientific endeavors including, but not limited to, natural history (and may include astronomical observation, as featured in book 17). That being said, Haywood does recreate within the pages of her polite essay periodical a kind of microcosm of naturalist study as exemplified by *Philosophical Transactions:* she embraces prevailing standards for scientific inquiry and observation but locates such inquiry in a domestic setting to make it more accessible to women. Furthermore, she leverages women's relationality to the domestic to generate for herself, her characters, and her readers a distinct kind of empiricism that has the potential—limited as it may be—to circumvent a patriarchal society and even prove men wrong once in a while. Thus, by linking the Female Spectator's endeavors to those of the Royal Society, Haywood reaffirms the grand potential of naturalist study for women; it can not only expand their intellectual horizons but even allow them to lay claim to a domestically rooted scientific authority.

It is not my intention to establish a sustained antagonism between the Female Spectator and Philo-Naturae, as that would misrepresent what Haywood is trying to do in these passages. Rather, I point to the slippage between the text of the Female Spectator and the text of the male Philo-Naturae as a space in which Haywood is able to articulate a specific model of studying natural history that exceeds conventional bounds of women's intellectual spheres while remaining feasible for her readers. Where the male correspondent presents reading and study as too abstruse and advances instead a theory of limited physical action, the female eidolon (already having endorsed the reading of natural history texts) pushes that action beyond expected boundaries yet maintains a limit within appropriate feminine decorum. The preservation of what might be considered appropriately feminine, I reiterate, is far from a confirmation of Haywood's purportedly conservative vision for women's lives. Instead, it is an acknowledgement of the very real limitations and challenges women faced in the eighteenth century and a hint about how women might circumvent such limitations in the pursuit of intellectual authority and personal satisfaction.

Charlotte Lennox's Strategies for Reading

Like Haywood, Charlotte Lennox was a prolific writer who wrote to support herself and her family for much of her life. Although over the course of a long career she wrote a number of novels (including the first serialized novel, *Sophia*) and original poetry; translated histories, plays, and memoirs; and even produced the laborious critical work *Shakespear Illustrated* (1753–54), Lennox is best known for the popular novel *The Female Quixote, or the Adventures of Arabella* (1752). Published approximately ten years prior to the *Lady's Museum*, *The Female Quixote* is a notable precursor to the strategies of reading and education she would employ in her periodical. Thus, I begin by examining the practices of reading and self-education that Lennox first articulates in *The Female Quixote* to provide a fuller context for her later periodical and how she approached scientific learning for women.

The Female Quixote has been conservatively interpreted as the story of a romance-addled heroine who ultimately comes to her senses after a series of dangerous incidents and meekly accepts her absorption into a conventional domestic novel. While most contemporary criticism of the novel attributes much more nuance to the story, I want to explore the radical potential behind the heroine Arabella's investment in romances. Despite the frequent laments of her cousin and prospective bridegroom Glanville, it seems evident that Arabella's good qualities are explicitly connected to her romance-reading habit.[57] The speech that most impresses her suitor Sir Charles is a lengthy disquisition against "raillery"—a subject upon which Arabella's views conform closely to the conservative stance on the virtue of "wit"—and is in fact drawn nearly directly from one of Arabella's favorite romances, the compendious *Artamenes*.[58] Unlike her conversation with the pretentious Selvin, during which the reader may chuckle knowingly at the labeling of Scudéry as a "historian," in this scene none of her listeners recognize these sentiments on raillery or their essential moral value as rooted in romance.[59] In fact, despite the narrator's representation of Arabella's ardent emulation of her heroines as only a love for "romantick Heroism," her earnest espousal of "Virtue, Courage, Generosity, and Honour" distinguishes her repeatedly from her shallow cousin Charlotte.[60] Sharon Smith Palo explains, "In the opinion of . . . educational writers, romances teach young women to revere admirable human qualities, improve women's conversation and their understanding of language, and provide knowledge of a world with which they might otherwise have limited experience."[61] By the mid-eighteenth century, this endorsement would have been somewhat less enthusiastic, qualified as beneficial only if framed within the appropriate historical context and consumed in moderation.[62] As the Countess explains to Arabella, "Tho' the Natures or Virtue or Vice cannot be changed . . . yet they may be mistaken; and different Principles, Customs, and Education, may probably change their Names, if not their Natures."[63] Scott Paul Gordon elaborates, "The countess, in effect, historicizes romance. . . . These things *happened*, rather than *happen*."[64]

What does Arabella learn, or think she learns, from her reading? Aside from her encyclopedic memory for grandiose love stories and tragic wars, Arabella also prides herself on a particular knowledge of geography, religion, and social customs of ancient cultures—all key areas of inquiry in an eighteenth-century educational curriculum for women. Indeed, Arabella seems to have an almost eidetic memory for the geography of the major Mediterranean nation-states in classical history, and it seems plausible that she is familiar with (although an unlikely reader of) Herodotus, Plutarch, and Thucydides, given that she responds without pause to Selvin's name-dropping.[65] Arabella's lecture on the Olympic Games, as well as her enthusiasm for their cultural and historical value, echoes what Margaret Dalziel identifies as information drawn directly from one of the most authoritative texts on the subject in the early eighteenth century, Temple Stanyan's *Grecian History* (1707).[66] Ruth Mack's suggestion that Lennox was influenced by Montesquieu's *The Spirit of the Laws* (1748) in writing some of her dialogue relocates Arabella's recounting of historical facts into the realm of historiography, as the heroine not only retells each story, but also interprets it for her listeners along the way.[67] Arabella's reading habits have allowed her to hold "two kinds of perspectives at once," to negotiate between them, and to make informed choices that consider multiple and sometimes contradictory accounts.[68]

In a sense, then, Arabella behaves like a historian or ethnographer, with critical abilities developed from her reading that exceed any such talents of her compatriots—male or female. And, although she misuses these critical reading skills for a time, there is no reason to suppose that she would not appropriately redirect those skills in the future.

Ultimately, the novel's frequent slippage between romance and history calls into question the possibility of emphatically endorsing one form of reading over another. For Arabella and the readers of *The Female Quixote*, the task seems to be not a matter of sorting out fiction from reality, but, as Mack puts it, understanding that "the contest for truth must be fought between texts."[69] An excellent example of Arabella's ability to negotiate multiple texts comes early in the novel, when she is forced to reconcile her romantic standards of behavior to the realities of eighteenth-century life. Having convinced herself that her cousin's clumsy attempts at courtship hinted at violent intentions, she is determined to run away but finds herself struck with a sudden incongruity: "The Want of a Precedent, indeed, for an Action of this Nature, held her a few Moments in Suspense; for she did not remember to have read of any Heroine that voluntarily left her Father's House ... but she considered, that there was not any of the Ladies in Romances, in the same Circumstances with herself who was without a favored Lover."[70] Equally aware of the impropriety of such a course of action in modern England and of the troubling "Want of Precedent" in her historical romances, Arabella is faced with the necessity of exercising her powers of critical interpretation. Palo suggests that this quandary provides an opportunity for her to "[demonstrate] her ability to think and act beyond the limitations imposed by the codes of female conduct available to her," and I would add that it also showcases her intellectual flexibility and critical thinking skills.[71] Obliged to acknowledge both romances and eighteenth-century codes of female conduct *as texts*, Arabella makes her decision based upon an integration of the two. Her father's sudden illness prevents her from enacting one of her "thousand different Stratagems" of escape, but it does not erase her demonstration of intellectual acuity.

The novel thus ultimately seems to suggest that most forms of women's education can be understood as a negotiation between texts, in which women must adapt their reading—of romance, natural history, geography, and biography—to the often-contradictory realities of their daily lives. The trick for women who aim to self-educate like Arabella is not to avoid the attractive narratives of heroic biography, travelogues, or geographic studies (all of which were commonly presented to women in the eighteenth century as appropriate subjects of study), or even the factually questionable but ultimately edifying romances, but to study each subject in conjunction with the others (like the Countess) and in the appropriate context—a context Lennox would go on to provide in the *Lady's Museum*.[72] And so, by 1760, Lennox was able to present a fully developed articulation of the appropriate tactics for self-education that she had explored through the character of Arabella a decade prior.

The *Lady's Museum* and Self-Education

The *Lady's Museum* was issued in eleven monthly numbers from March 1, 1760 to January 1, 1761, and was later published as two collected volumes for a total of approximately eight hundred manuscript pages of text.[73] At a modest cost of one shilling, the single issues would have been affordable for women of the middle class (who likely made up the bulk of Lennox's readership).[74] The bound collections would have been less so, although, as Ros Ballaster notes, "with the establishment of lending libraries in the mid eighteenth century, those who could not afford the full cost of the magazine could borrow it for a minimal sum."[75] The *Lady's Museum* was credited to "the author of *The Female Quixote*" on individual issues and was directly attributed to Lennox in an advertisement in the *London Chronicle*.[76] Each issue began with a brief topical essay under the heading "The Trifler," the young female writer of which may be said to be the only real editorial voice in the publication, and continued with a selection of other, longer pieces.[77] In addition to essays on natural history, geography, and astronomy, the *Lady's Museum* also included a number of translations, selections of Lennox's poetry and fiction (some of which had been previously published elsewhere), and the serialization of *Sophia*, under the title "The History of Harriot and Sophia."

Broadly speaking, Lennox's publication was a miscellany periodical for women, albeit one with rather longer excerpts and fewer entries per issue than was typical of that format.[78] Lennox fortunately made significant connections in the print industry—including her husband's sometime employer, printer William Strahan, and her close friends Samuel Johnson and Samuel Richardson—although the extent to which these men contributed to advertising this project is unknown, due to a gap in the extant correspondence from the 1760s.[79] The *Lady's Museum* was positively reviewed, and, like many periodicals "for the Ladies," was considered edifying tea-table reading for women of diverse classes and occupations. However, it sold modestly and was discontinued, likely due to lack of profit and Lennox's own ill health.[80]

While most critics who address Lennox's periodical acknowledge that the *Lady's Museum* possessed pedagogical elements, women's education is often considered a subsidiary interest in the text (usually read as ancillary to Lennox's financial motives). Developing critical work by Iona Italia, Susan Carlile, and Claire Boulard Jouslin (as well as my own work on the historical pieces in the *Lady's Museum*), I argue that the pedagogical imperative of the periodical was Lennox's primary goal. In the first prefatory "Trifler" essay, Lennox identifies an expressly female audience for the periodical: "Universally as I could wish to please in this paper, yet I shall be contented, if it finds only a favourable acceptance with my own sex, to whose amusement it is chiefly designed to contribute."[81] She clarifies that the focus is not only amusement, but also education, as subsequent essays with titles like "Of the Studies Proper for Women," "On the Importance of the Education of Daughters," "Geography for the Ladies," and "Philosophy for the Ladies" indicate.

In fact, "Of the Studies Proper for Women," a translation of Pierre Joseph Bourdier de Villemert's *L'ami des femmes* (1758) that Lennox completed herself, is the very first article after the Trifler introduction, attesting to the importance of its subject matter.[82] While Villemert's text would only appear in this first issue, François de Salignac de La Mothe Fénelon's *Traite de l'Education des Filles* would be featured in significant excerpts in nearly every subsequent issue, forming what Jouslin argues is "the periodical's backbone."[83] This volume was especially popular among the abundance of pedagogical texts aimed at women in the seventeenth and eighteenth centuries; it was also an immensely profitable educational treatise to reproduce. First published in France in 1687, *Traite de l'Education des Filles* was translated into English in 1699 and again in 1707 in an edition by George Hickes entitled *Instructions for the Education of a Daughter* that was reissued seven times before 1750 alone.[84] Avowedly pedagogical texts thereby populated the *Lady's Museum* from the get-go, further underscoring Lennox's investment in imbricating edification with entertainment throughout.

However, if "Of the Studies Proper for Women" (a rather conservative text) were to be taken as the foundational philosophy of the *Lady's Museum*, one would consequently expect certain other elements common to miscellany publications that were typically thought to "improve [a woman's mind]": brief reports of contemporary events, important political developments, carefully curated popular poetry or fiction, advertisements for trade goods or products, recipes, clothing patterns, or even scientific and mathematical question-and-answer sections.[85] While the *Lady's Museum* did include some of those elements—the reprinted poetry and serialized fiction featured quite prominently—it also crucially omitted others that were thought to facilitate polite conversation in mixed gatherings.[86] For example, Lennox included only one entry on a current event, a letter in issue four from an anonymous correspondent recounting the trial and execution of Laurence Shirley, 4th Earl Ferrers, for murder—an event that would have been sufficiently notorious to merit inclusion despite a standing policy to exclude news.

Typical content was less salacious, however. In addition to historical and biographical information excerpted at much greater length than was common, Lennox included a glowing essay on the personal benefits of studying natural history (an area of inquiry perfectly suited to "satisfy and furnish food" for a woman's curiosity—a curiously carnal descriptor), a charming rhapsody on astronomy, and a detailed geographic and cultural study of the islands of Amboyna and Ceylon, including such specificities as diet, dress, and social customs.[87] Thus, as Jouslin explains, far from duplicating the recommended curricula of Villemert and Fénelon, "Lennox . . . chose through the other parts of the periodical to define female learning as one that provided girls with intellectual knowledge that was not limited to practical domestic application."[88] Kathryn Shevelow points to a similar discursive phenomenon in magazines that encapsulates Lennox's strategy, in which the ideological thrust of the publication was generated not by a singular authority, but by an accumulation of "various different types of expression."[89]

The model of self-education in the *Lady's Museum* is thus partially contingent upon the format in which the material was presented. Just as the liberal miscellany of content allowed for greater freedom of interpretation, the mixture of content and style offered a rare element of personal choice. That is, the periodical employs a digest model of education in which women have control over the material they choose to read, while at the same time presenting the material in such a way that invites critical comparisons between its discrete components. By including a detailed description of the geography of Ceylon adjacent to an excerpt from *Orlando Furioso* and a historico-romantic biography, "The Life of Sir Anthony Vandyck," Lennox offers each as equally valid targets of study. Furthermore, without the overt commentary of the eidolon in the style of the *Female Spectator*, the periodical's content speaks for itself and can be studied by the reader on her own terms—as single entries, as one of many subjects in a single issue, or as an entry in an ongoing topical series.

The *Lady's Museum* thereby offers a distinct model of learning for women, in which Lennox provides the opportunity for women to absorb detailed knowledge in the presentation of accessible material that "exercises its liberty of vision with a startling contraction of perspective."[90] The especially poetic quality of many entries was also likely to appeal to women who, like Arabella, might have been more accustomed to reading romances than textbooks. In at least one sense, then, Lennox does concur with Fénelon: education is best grounded in the reading of books, particularly those "that shall contain a number of stories, short and surprising" because "whatever delights the fancy, is an assistance to study."[91] The effectiveness of self-education, Lennox agrees, is inextricable from the appeal of the format—after all, what is the *Lady's Museum* in most basic terms but a collection of "stories, short and surprising"?

A Scientific Digest

History is only one part of the educational material included in the *Lady's Museum*.[92] In particular, Lennox especially encouraged "a curious inquisition into the wonders of creation" as an appropriate intellectual pursuit for women.[93] The scientific components of the periodical are the "Philosophy for the Ladies" series, "The Lady's Geography" series, and the singular "Of the Universe, as Considered under a General View," which broaches the study of astronomy. On these subjects, Lennox (viz Villemert) recommends a cautious, measured approach that initially seems suggestive of limited feminine intellectual capacity: "I would particularly recommend to [women] to avoid all abstract learning, all thorny researches, which may blunt the finer edge of their wit and change the delicacy in which they excel into pedantic coarseness."[94] Fénelon agrees—"It is true that we ought to be very cautious of making pedantick ladies"—but does Lennox?[95]

Reconfigure this objection in the context of the educational material offered in the scientific entries themselves and it becomes clear that Lennox does not express sincere distaste at the unpleasant thought of an overeducated woman. To the contrary, Lennox's model of education in the *Lady's Museum* seeks to renovate scientific reading practiced by women in such a way that suits the study to their gender. Consequently, the pejorative epithet "pedantick" might be interpreted in Lennox's periodical more as the result of ill-fitting match between a woman reader and the material she chooses to read, rather than a quality of being especially learned. Thus, Lennox proposes a revision of the means by which women study natural history: "If therefore we treat of philosophy, it shall be polished from the rust of theoretical erudition, and adorned with all those advantages which a connexion with the politer arts and sciences can throw upon it."[96] Women who are exposed to scientific subjects in the periodical are not engaging with an inferior articulation of the subject (as contemporaries would have it), but, according to Lennox, are pursing the study in a manner more amenable to their style of learning.

As a much less expensive and more easily obtainable text than a technical scientific work, an already gender-appropriate periodical like the *Lady's Museum* would have been a practical option for educating young women not only on natural history, but also on a variety of subjects. Periodical issues were designed for brief periods of reading rather than intense study, and reading small excerpts of historical or scientific material would have been possible in many households; they were lighter than bound books, and although the individual issues of the periodicals could be bound for preservation, the individual issues were eminently portable and easily shared. Given that reading (both to oneself or aloud to a companion) while doing domestic tasks such as mending or tatting was common, the educational excerpts in the *Lady's Museum* would have been ideally suited for many young women's reading practices and were likely printed with them in mind. Consequently, Lennox's use of the periodical form as her pedagogical tool (rather than, say, a separately published critical study, like her *Shakespear Illustrated*) was a canny decision.

Like Haywood, Lennox is careful to address the issue of utility with regard to reading natural history and astronomy. In introducing "The Universe, as Considered under a General View," she insists that "every study ought to have its particular use not only with regard to mankind in general, but to the person to whom it is recommended in particular."[97] While she elaborates upon the inevitable links between intellectual discovery and ambition, she implies that the usefulness of any dedicated study for women lies primarily in the spiritually soothing effect it incites; such studies "lead by smooth and regular gradations to peace and happiness: they raise the thoughts to humanity and devotion; serve to calm our ruffled passions, and, by a regular transition, convey our contemplations from the creature to its Creator."[98] And so the text initially suggests that the study of subjects like astronomy, while enjoyable, are to be undertaken primarily for their spiritual benefits rather than to satisfy any kind of personal intellectual thirst. However, this conservative appraisal comes after the following emphatic endorsement of scientific study: "The so common cry against the practice of natural philosophy, What is the use of this? A cry thus raised as easily is answered, The use is universal."[99] It seems evident by comparison that Lennox's endorsement of scientific pursuits is framed by that artificial caveat; juxtaposed with a moment of true passion, the claim of a "universal" use is gradually spun out across the periodical's varied entries and issues.

On the whole, the *Lady's Museum* presents the subject of natural history rather differently than the *Female Spectator* does. To some extent, this is a product of the fact that Lennox's publication coincides with a significant shift in the popularity of subordinate areas of inquiry like botany, following the translations of Linnaeus's landmark works in the 1760s. Although Lennox's work with the *Lady's Museum* was published on the cusp of this shift in scientific interests, it is worth noting the difference twenty years makes in the climate of popular opinion: whereas Haywood's audience was avidly interested in the microscopic studies of insects and other organic material that Hooke made popular in his *Micrographia*, by the second half of the century the popular conception of natural history seemed narrowed into "botanomania." For women in particular, an almost exclusive focus on vegetal subjects was endorsed as both more appropriate and accessible; by 1799 Lady Charlotte Murray would note in *The British Garden* that "whilst the research into the animal kingdom is attended with many obstacles which prevents its general adoption, and preclude minute investigation . . . the study of Botany . . . is open to almost every curious mind."[100] While this did not preclude entomological subjects from attracting a robust reading audience, it could have contributed to some of Lennox's trouble in finding a sustainable audience for the periodical. On the other hand, changes in format, layout, and style in post-Linnaean publications may have been conducive to the flourishing of alternative models of scientific pedagogy. Practically speaking, the spread of this Linnaean system gave rise to an encyclopedic model of presenting scientific information to women, a model that Lennox effectively took advantage of in her deployment of scientific material in the miscellany format.

Lennox broaches the subjects of natural history or astronomy in seven of the eleven published issues and, in those issues, links it overtly to her avowed pedagogical manifesto. In keeping with the prevalent trends in textbooks of natural history that especially characterized the later decades of the eighteenth century, she presents the subject through a series of narrative entries.[101] Somewhat atypically for naturalist texts written for women in the 1760s, Lennox focuses primarily on entomological or zoological subjects, avoiding altogether the descriptions of common plants and flowers that filled most popular scientific texts of the decade. Instead, she brings her readers' attention to the power and beauty of the small and insignificant creatures in nature, like the swallowtail butterfly or the dayfly.[102] While we cannot know if she intended to cover botanical subjects in subsequent, unwritten issues, the information she does provide in each edition of "Philosophy for the Ladies" makes clear that the purpose of natural history is to entice women into more abstract, profound contemplation.

Like Haywood, Lennox links scientific study to the domestic space of the garden and the home. Both more approachable and more accessible than obtuse scientific tracts, her miniessays, explicitly "for the Ladies," offer the information in a model of reading they are already accustomed to. The creatures to which she devotes her various issues—the ephemeron, the swallowtail butterfly, caterpillars, even rabbits and foxes—are quotidian and recognizable, either found in England or akin to a similar species found somewhere in the British Isles. While offering the promise of more exotic creatures (discussing in some issues, for example, insects from New Zealand and small mammals from the Arctic Circle), Lennox confines her concrete information to subjects with which her readers are likely already familiar. Indeed, the enticement to study natural history in the *Lady's Museum* is very often rooted in the allure of discovering something new in the everyday; the featured objects of study are unlikely to incite any censure and can be found in and around a woman's home, rendering the proposed scope of study one rooted firmly in domestic ideologies and spaces.

Part of this emphasis on domesticity meant that the *Lady's Museum* avoided the evocation of peripheral authorities like popular textbooks or optical technology. Instead, Lennox introduces natural history to her readers through an engaging presentation of certain topics herself. For example, she writes eloquently and in great detail about the varying developmental stages of certain mammals and the differentiation between mammalian reproductive categories, despite the prefatory caveat that "[her] wish is to render the ladies though learned not pedantic [and] conversable rather than scientific."[103] In fact, she devotes several pages to a thorough narrative of the life cycle of the swallowtail butterfly, covering everything from the dietary proclivities of the caterpillar to the typical arrangement of its eggs on a given plant, alongside attractive illustrations. Drawn with a level of sophistication and accuracy that rivals similar illustrations from scientific textbooks, these images and the essay they accompany turn the pages of Lennox's periodical into a sort of naturalist textbook. In this respect, Lennox was also likely taking advantage of the substantial market interest in illustrated naturalist books to sell copies of her periodical.

The subjects of these essays are especially germane to an understanding of Lennox's strategy for connecting women and nature. Like Haywood, she is interested in insects commonly affiliated with domestic spaces; but unlike her predecessor, Lennox does not prevaricate in her claims to scientific authority. Her essay on the indigenous swallowtail butterfly, for example, focuses extensively on the metamorphosis undergone by the caterpillar and its various stages of embodiment. She begins by praising this species as an appropriate subject of study for women because of its appearance and the moral benefit that studying its transitions offers (having already explained the concepts of rebirth, metamorphoses, and metempsychosis in a previous issue).[104] Lennox writes: "That beautiful and almost infinitely varied genus of insects, which is so well known by the names of butterfly and moth . . . is to be ranked with those of the most perfect change, as every one of this kind passes through the greatest number of states that has yet been discovered amongst animals, viz. the egg, worm, chrysalis, and fly."[105] Several following pages cover everything from the typical number of eggs laid by an adult butterfly (between two and four) to the caterpillar's favorite food (fennel and other umbellate plants). Unlike Haywood, Lennox does not dwell on the particular beauties of the caterpillar, preferring to describe it in more technical terms, evoking Linnaean natural history both in name and in approach: "Its body is green, annulated, or striped transversely with striae of black, each ringlet studded with spots of a bright scarlet."[106] Such descriptions nearly identically echo the highly technical language found in George Shaw's *Naturalists' Miscellany* (1789), which, although published much later than Lennox's periodical, is an exemplary version of both style and substance in eighteenth-century naturalist tomes. Lennox's further discussion of the insect's lifecycle is clinical, as she moves easily through the stages of the *Papilio machaon*'s development with little extraneous commentary or poetic rhapsodies of the type that she would later employ with her discussions of astronomy. The text even dispassionately details a common, albeit incredibly violent disruption of the transition from larva to pupa: the attack of the parasitical ichneumon.[107]

This lack of empathy places Lennox in a position quite different from that of the Female Spectator. While the latter gushes over the beauty of a caterpillar and lovingly cares for a coterie of immature snails, the former transmits information with little annotation. What is notable about Lennox's narrative strategy for studying these domestic creatures, however, is the way she conveys an enthusiasm for technical minutiae. Far from shying away from the more complex details of insect reproduction (or mammalian reproduction, found in a section on viviparous animals), she provides for her female readers an undiluted flow of scientific information.[108] Rather than offering the radical possibility of women identifying with these fascinating creatures, Lennox instead offers the equally daring potential for women to become as expert in naturalist subjects as their male counterparts. The relationship between women and insect here is not one of affiliation, but it is not one of disjuncture, either. Female bodies devote themselves to studying insect bodies through the pages of Lennox's periodical—an intermediary, certainly, but an active and invested one.

The *Lady's Museum* also serves as an emissary to the field of astronomical research, introducing the subject as one of unblemished pleasure and wonder. Lennox initially reminds readers of the necessity "to view at once with wonder, and follow with discernment, the motions of heavenly bodies" in any kind of astronomical study.[109] A longer passage evokes the kind of visceral fantasy favored by Arabella, in which Lennox beguiles her readers to imagine an out-of-body astral experience: "Reader, imagine yourself conveyed to some place beyond even the limits of infinite space; there cast your eyes around, and view the number of the stars which glitter in their several orbs. Small as they from our earth appear, behold them each a sun, shining with brightest lustre; each an immense mass of heat and light. Around them see numbers of worlds revolving . . . lost as you are in the irregularity of their motions, now fix your attention on a single one—return to our system only."[110] Lennox continues to use this device of imaginative transportation throughout the essay, while punctuating her descriptive language with ecstatic asides ("How wonderful a combination! how inconceivable to human fancy, the impulse by which it could be at first set to work!") to repeatedly reinforce the intrinsic connection between pleasure and scientific study.[111]

Yet instead of stopping at these moments of awe, Lennox uses the attractiveness of rapturous language to introduce her readers to technical information about the universe; leading their imaginations on a merry dance between the planets, all the while providing specific details like Mercury's distance from the sun and length of solar year, the phases of

Venus, and the Moon's orbital axis.[112] As she explains, her pedagogical strategy is one of deliberate incrementation: "Look on the vast universe as one immense machine. . . . First take a view of this mighty machine in the whole, and then descend to a more immediate disquisition of its several parts."[113] Thoroughly empirical and yet thoroughly imaginative, Lennox's model of scientific study in the periodical represents an evolution of earlier feminist strategies for self-education like those of Haywood in the *Female Spectator.*

Lennox's scientific writings in the *Lady's Museum* also encompass an impressive diversity of subjects that exceeds the simple topics conventionally presented to female readers, making her periodical more daring than most. She declares in an early issue, "One Number of our work perhaps shall leave us admiring the stupendous fabric of the immense extended universe; the next shall find us aiding our limited sight by help of glasses in observations on a world of unknown beings within a drop of fluid, or forests waving in the narrow circuit of a small piece of moss."[114] The almost unimaginable variety of subject matter in the *Lady's Museum*—a reading curriculum that spans the far reaches of "the immense universe" to the invisible world of bacteria "within a drop of fluid"—is rendered approachable through Lennox's decision to present the scientific information in a top-down model. Each issue develops upon the last, continuing in a series of brief but densely informative additions to the larger entry on the given subject. Collected together, the issues contain almost encyclopedic specificity; read separately, they do precisely what Lennox claims: enlighten without overwhelming. With such an observation in mind, Susan Carlile's description of the *Lady's Museum* as a kind of virtual salon seems especially apt.[115] Unfortunately, the project was discontinued before the full range of Lennox's ambitious plan for her readers was fully realized.

It is clear that Lennox's pedagogical strategy aimed for far more than the accessibility of scientific material. The tiny subjects of the scientific articles, as enlightening as the description of their transformations may be, also emphasize a conceptual alignment between women and the more insignificant creatures in life—a pathos Lennox acknowledges but ultimately uses to her critical advantage. In the last issue of the *Lady's Museum,* Lennox beckons her readers to "mark the light, the tender, the seemingly insignificant ephemeron, with a life designed for but for some hours continuance, burst from its embrio [sic] state in one element, and almost imperceptibly become the inhabitant of another, enabled to rove unlimited, and taste of every pleasure his being will admit of."[116] Like the dayfly, "seemingly insignificant," but unlike the insect, without the freedom to "rove unlimited," women often find their lives constrained by decorum or by circumstance, prevented from experiencing an intellectually satisfying life. Yet "Philosophy for the Ladies" offers a promise alongside this somber reminder. Like Arabella, Lennox's readers can take pleasure in the worlds to which their reading brings them; through the pages of the periodical, readers of the *Lady's Museum* actually can "taste of every pleasure" and "rove unlimited" through the stars above and the waters below. It is only in her final issue that Lennox achieves the kind of interspecies sympathy that Haywood cultivates in the *Female Spectator*, but it is time enough to establish the significance of such intimate connections between women and their scientific subjects.

Lennox continues to encourage her readers to draw hope from the lessons learned through this study of natural history and astronomy and to find comfort, perhaps empowerment, through understanding their place not merely in the context of English society, but as a part of a great and majestic universe. She explains, "Every one of the organs of this grand machine, not only the larger and more apparently useful, but even the most minute, insignificant, and almost invisible ones, are furnished in the amplest manner, not barely with such parts . . . as are needful for their mere existence, but still more particularly with such peculiar contrivances, such sagacity, such intellectual faculties, as must render that existence, with respect to the place, station, and allotment of each individual, absolutely and perfectly happy."[117] Although at first glance this passage seems reminiscent of the more conservative assessments of a woman's role in society and the home, Lennox emphasizes the potential for personal fulfillment and a sense of self-worth over the restrictive notion of an inflexible "place, station, and allotment." This has been Lennox's educational philosophy all along. Not, in the manner of Mary Wollstonecraft or Mary Astell, a radical reenvisioning of a woman's place in society, nor, as her detractors would have it, a regurgitation of conservative limitations on feminine education, Lennox's innovative pedagogy encourages each woman reader to adapt the materials available to her, like a miscellany periodical, to a style of learning that fits her intellectual abilities and her lifestyle. While one woman may read about the ephemeron in an issue of the *Lady's Museum* and be inspired to seek out further information in popularizing textbooks, another may simply take time to read an individual issue and brighten up the day by learning something new. Always aware of the social and financial limitations on women in mid-century England, Lennox's model of education in the *Lady's Museum* aims to work within those limitations and to console women with the promise of marvelous things to come.

The *Lady's Museum* enticingly presents the subject of natural history as a potentially unending source of enlightenment, wonder, and personal satisfaction. Lennox grandly proclaims, "Even to the farthest stretch of time, our researches into the wonders of Nature's inexhaustible storehouse, would have been no other than the pursuance of an apparent horizon, the boundaries of which are ever flying before us, and although they every moment present us with a fresh variety of enchanting objects, yet are, with respect to ourselves, as absolutely distant at the last as the first moment of our journey."[118] Limited though each discrete session of reading "Natural Philosophy for the Ladies" may be, the subject itself is boundless. The "fresh variety of enchanting objects" is always available for women to study; a completely appropriate topic linked to both spiritual enlightenment and practical education, natural history has the enticing benefit of pleasure mixed with that utility. In fact, even the smallest moments of observing the natural world can bring joy, as she advises her reader to "observe the little duckling bursting from the egg, and rushing instantly into an unruly, a destructive element, to pick up food, and taste the joys of living." Scientific study in the *Lady's Museum* is ultimately empowering; made as attractive as a romance and as socially acceptable as a volume of history, natural history in this periodical is framed to suit a woman's tastes and her abilities, without any genuine intellectual dilution or restriction.

Reading the Periodical as a Scientific Form

I suggested in the beginning of this chapter that the periodical genre possessed an inherent fluidity that other texts in the eighteenth century did not, and the rich field of eighteenth-century periodical studies has demonstrated the vast potential for mutability in this most hybrid of genres. With its eidolons, contributors, and cast of characters real and imagined, periodical literature offers important insights into the construction of authorship, and, more pertinently, the construction of intellectual authority as an example for women. For both Haywood and Lennox, the periodical form offered significant opportunities for promoting female scientific learning.

Haywood took canny advantage of the versatile possibilities of reading offered by the periodical, a distinction that allows us to see Haywood's project for what it really is: not merely one among a plethora of magazines for "Ladies," a latecomer to the tradition of essay periodicals, or even the bearer of the interesting but ultimately irrelevant honor of being "the first magazine for women written by a woman."[119] Rather, the *Female Spectator* deftly combines a number of discursive strategies, textual patterns, and recognizable stock characters from Haywood's periodical milieu into what can ultimately be read as one of the more innovative presentations of women's scientific education within the genre. For her own part, Lennox adapts a more conventionally feminine periodical format, the miscellany, into her reconfiguration of strategies for reading scientific subjects. Keenly aware of the potentially limited opportunities for reading available to her readers and equally cognizant of the critical fear of "pedantick ladies," she contextualizes an empirical construction of scientific knowledge amid romantic histories and polite essays.

Haywood and Lennox appear to take comparable, although importantly different, stances on the subject of natural history. Both authors strategically root the subject in domestic spaces and quotidian subjects, and both periodicals consistently emphasize the moral and spiritual value of contemplating nature. Lennox at first glance would appear to endorse the subject more fully than Haywood; the *Lady's Museum* presents a level of technical detail and corresponding illustrations that match popular textbooks and focuses more specifically on ideas like mammalian reproduction and entomological lifecycles. On the other hand, the subject of natural history in the *Female Spectator* is endorsed by male correspondents and consistently edited to avoid too much technical detail. Yet in her own way each author offers distinct, equally enlightening practices of reading natural history that depend upon the form of the periodical for their successful execution. While Lennox presents the study via a series of small, digestible sections through which practically limitless subjects can be methodically tackled, the *Female Spectator* encourages readers to go out and make new discoveries themselves. Haywood's earlier publication, with its roots in the single-essay periodical and the imaginative cast of eidolon characters, can present scenarios in which women readers envision themselves alongside Mira and the Female Spectator, tending snails in a flower pot and holding caterpillars under a microscope. Lennox's miscellany structure renders a difficult subject more approachable through the twin strategies of contextualizing scientific information alongside history and romances, and breaking large subjects into discrete, digestible lessons.

The practice of natural history in the world of the *Female Spectator* and the *Lady's Museum* is not a revolutionary proposition, but that is precisely the point. Scientific study will not take a woman away from the responsibilities and social sphere to which society confined her; it will not defeminize her; it will not segregate her from company nor make her disagreeable to her husband—all factors that both authors carefully emphasize. Haywood explains, "All that was aimed at in giving this Account of what little Observations we were able to make, in our short Excursion from *London*, was to shew the Female Subscribers and Encouragers of this Undertaking, how much Pleasure, as well as Improvement, would accrue to them by giving some few Hours, out of the many which they have to spare, to the Study of Natural Philosophy."[120] Yet this hedging framework—"all that was aimed"—has been clearly undermined by frequent moments of pleasure and genuine enlightenment that both Haywood and Lennox present in their depiction of the subject. Pleasure and improvement are not idle things, but sources of true satisfaction and fulfillment for a woman living in the patriarchal mid-eighteenth century. Ultimately, reading natural history can be a woman's way of ameliorating the troubles and the restrictions of her daily sphere; it has value not only in the moment of practice but as an enduring presence that can continue to brighten and enhance a woman's life.

Likewise, the study and practice of natural history as specifically rooted in the domestic space of the garden offers opportunities for women as writers, readers, and naturalists. By grounding their advocacy of natural history in a simultaneously subversive and decorous space like the garden, both Haywood and Lennox take advantage of the productive mutability of such a space to envision new ways of being for women. With the endless possibilities contained within the study of naturalist subjects, women in the eighteenth century could find in these works the freedom and flexibility to "take a view of Heaven itself, and traverse all the Wonders of the Skies," as Haywood promised. While such movement did not always entail literal journeys, the imaginative adventures undertaken by women readers of the *Female Spectator* and the *Lady's Museum* and the shared experience each of these readers enjoyed surely contributed to the sense of community among women which was so valuable in the period. Further, by taking the advice to get out in the garden, track down specimens, and design their own experiments, readers could create for themselves a world in which they, too, could actively engage in and create scientific content. Finally, by celebrating the connections between women, gardens, and their scientific subjects, these periodicals cultivated a priceless culture of domestic scientific practice. And, where these periodicals offered casual exposure to naturalist subjects, as I explore in the next chapter, textbooks provided opportunities for interested women to pursue more profound studies of plant life and to further negotiate their own relationship to the natural world.

2

From Native Blooms to Monster Plants

Women's Botanical Textbooks and the Lives of Plants

The conduct and manners of women, in fact, evidently prove that their minds are not in a healthy state; for, like flowers which are planted in too rich a soil, strength and usefulness are sacrificed to beauty; and the flaunting leaves, after having pleased the fastidious eye, fade, disregarded on the stalk, long before the season when they ought to have arrived at maturity.

—Mary Wollstonecraft, *A Vindication of the Rights of Woman*

MONSTROUS Flowers, such as the *Multiplicate, Full,* or *Proliferous,* derive their Origin from natural ones, and therefore are to be considered only as a Variety from *Luxuriance.*

—James Lee, *Introduction to Botany*

As demonstrated by the periodical publications of Eliza Haywood and Charlotte Lennox, women were actively involved in practicing and publicizing naturalist labor in the mid-1700s, a trend that continued in the last years of the century as several women rose to prominence as textbook authors—none more significant than Maria Jacson and Priscilla Wakefield. While their texts were primarily intended to educate women and children in botanical science, their most famous works also reimagined the relationship between women and plants. In Wakefield's *An Introduction to Botany* (1796) and Jacson's *Botanical Dialogues* (1797), these authors revised the cultural parallels between femininity and vegetality as they privileged certain kinds of plant life (native and acclimatized) alongside specific female characteristics (resilient, intelligent, and strong) over detrimental floral metaphors like the kind Mary Wollstonecraft utilized to great dramatic effect. To a lesser extent, these texts also explored the connections between plants, vegetal monstrosity, and the human body, creating a world in which humans and plants might cultivate new connections. In doing so, they challenged dominant stereotypes about women and botany as well as prevailing ideas about the relationship of humanity to vegetality in the broader English culture.

When Wollstonecraft opened her polemical *A Vindication of the Rights of Woman* (1792) with the above analogy tying women to exotic flowers, her readers were familiar with the multivalent associations embedded in such a linkage. Evoking more than the superficial comparison between beautiful women and beautiful blossoms, however, the association of women with plant life ran deeper in eighteenth-century cultural consciousness. Linnaeus himself was fond of the analogies between women and plants, as evidenced in his lectures, published treatises, and various notebooks. In a journal entry from his trip through Lapland in 1732, for example, he describes the bog rosemary (*Andromeda polifolia*) as a virginal, blushing girl, declaring, "Scarcely any painter's art can so happily imitate the beauty of a fine female complexion; still less could any artificial colour upon the face itself bear a comparison with this lovely blossom."[1] Alan Bewell elaborates on such connections between femininity and plant life in the writings of botany's most famous taxonomist by explaining that, in the eighteenth century, "to see a plant properly . . . requires a special kind of double vision, which allows us to see both plant and female as one, passing easily from one to the other."[2] For Linnaeus and his fellow botanists, the boundaries between femininity and plant-being were both culturally and physically blurred; an ideal plant and an ideal woman shared much in common.

While men were the predominant purveyors of such imagery, women were certainly aware of the obstacles—both symbolic and tangible—with which they were faced in pursuing botanical study; as Theresa Kelley notes, "Women who wrote about or illustrated plants during the romantic era did so . . . at a time when the relation between women and plants was understood to be ornamental."[3] According to the eighteenth-century botanical imagination, women, like plants, were designed for show or display, intended for the superficial, rather than the profound. Furthermore, by the middle of the eighteenth century, the study of plants was also inextricably a study of sex, given the new emphasis on the Linnaean sexual system of plant reproduction. This analogy between women and flowers was far from complimentary: instead, it reduced women to bodies, sexualized, submissive, and vulnerable to "gathering." It was this that Wollstonecraft so strongly objected to: women-as-plants become weakened, ineffectual, by the loss of potential agency and the subsequent dependency upon others (specifically men) to survive. In using the metaphor of "too rich a soil" to refer to the idea of spoiled or effete women, Wollstonecraft not only tapped into a material history of women as things, but also a tradition in which the passivity of plants and their sexualized nature linked plant life to womankind in an entirely different way. In expanding the dimensions of the relationship between women and plants, she laid what would become the foundation for women like Jacson and Wakefield to advance in their work an innovative model of human-plant dynamics.

Julian Offray de La Mettrie, for one, had literalized this analogy earlier in the century in his botanical satire *L'Homme Plante* (1748). His observations paralleling women's bodies with plants are almost entirely sexual and meant to be titillating, comparing a "stylus" to "[a woman's] vagina; her vulva, and her mons veneris" and the "odour given off by the glands in these parts" corresponding with the "stigma."[4] Satirizing both the resonance between female bodies and plant bodies and the entire sexual system of botanical reproduction, *L'Homme Plante* represents only one small part of a larger tradition in which women's similarity to flowers was taken as a cultural given. After an extensive comparison between plants and women, detailing the parallels between body parts, smells, and sensations, La Mettrie concludes, "I have described botanically the most beautiful plant of our species, by which I mean woman. If she is virtuous, although metamorphosed into a flower, she will not be any the easier to gather."[5] Here, women are both culturally *and* physiologically plants, with the comparison of body part to plant part affirming a taxonomic assumption that Linnaeus and his predecessors made readily. However, I want to explore the ways in which women authors counteracted this sexualized floral metaphor by pointing specifically to work by Wakefield and Jacson that compellingly advances a different critical project while simultaneously retooling the affiliations between human life and plant life. Rather than replicating with slight difference the same tired concept of "women-as-flowers," these two authors reimagine how expansive and comprehensive the concept of "women-as-plants" can effectively be.

The relationship between human life and plant life has long been a complex one. As Londa Schiebinger explains, it was precisely this mirroring of plant and animal bodies that led to the development of the sexual system of plant reproduction: "It is clear that the recognition of sexual reproduction in plants was pushed by developments in zoology. . . . Botanists, committed to finding uniform laws of nature, increasingly drew analogies between plant and animal life, and thus transferred their interest in animal sexuality to plants."[6] Many of Linnaeus's predecessors utilized the parallels between plants and animals in their lectures and published works, and Linnaeus himself also wrote a (far less successful) system of taxonomy for animals in *Systema Naturae* (1735), borne out of the notion that principles of classification and division could be equally imposed upon theoretically comparable categories of life.[7] The idea that plants had analogous components to animals (specifically people) was something that women authors recognized as well; Wakefield and Jacson toyed with this idea in their botanical textbooks.

The fact that Wollstonecraft seems to reject so stridently the association between women and exotic plants leads most critics to rightly deduce that she was interested in endorsing a model of womanhood that rejected any intimation of dependency or weakness, and, in particular, sexual vulnerability. As Sam George argues, Wollstonecraft's "rationalist antipastoralism enables her to unmask the false sentiment behind the traditional metaphoric association of women and flowers and explode the familiar tropes, revealing them as embodiments of male desire, indicative of women's problematic relationship to culture."[8] While it is evident that the association between women and flowers was in large part generated by social and structural desires to keep women submissive and useful to men (i.e., beautiful and fecund), reading this passage a little more deeply—particularly in the broader context of women and botanical study in the late eighteenth century—yields compelling results.

Upon further consideration, the claim in this passage from *Vindication* is not so much that women could not be plants, or that women should not be plants, but that women should not be a particular type of plant. In other words, women should not be luxuriant, cultivated hothouse plants, useless outside of cosseted conditions and extremely susceptible to "plucking." For Wollstonecraft, botany as a field of study was actually quite significant in the context of the radical sexual politics of the 1790s; as Deidre Lynch puts it, "The sexual legibility of botany could be emancipatory for young women who, in studying plants, would better know their own physicality."[9] While this is precisely what angered Richard Polwhele and the other anti-Jacobin "protectors of female modesty," it is also what enables the complication of the metaphorical relationships between women and plants at the end of the century.[10] In fact, what Wollstonecraft wanted to accomplish (along with Wakefield and Jacson, as I will elaborate below) is a transition from one set of floral metaphors, of women as listless exotics or cosseted luxuriants, to another, more generative set of botanical metaphors, of women as hardy, native plants.

Wollstonecraft's continued scorn for floral metaphors can be seen in a later footnote in *Vindication,* in which she excoriates Anna Laetitia Barbauld for her poem "To a Lady, with Some Painted Flowers," which celebrates "flowers sweet, and gay, and delicate like you,"—another clear moment where "delicacy" is equated, for Wollstonecraft, with weakness rather than sensibility or another admirable female quality.[11] As the poem continues, what we can easily read as a derogatory connection between women and flowers becomes seemingly absurd:

> But this soft family, to cares unknown,
> Were born for pleasure and delight alone.
> Gay without toil, and lovely without art.[12]

The notion that women "were born for pleasure and delight alone" and should have neither "cares" nor "toil" seems to relegate women to a kind of slavish existence, where their sole purpose is to please—and to do so, of course, "without art," making their beauty and perfection seem effortless. A flower like this, according to Wollstonecraft, is a terrible model for femininity. As she argues throughout the text, Wollstonecraft is concerned with the potential for women to transcend this perception of themselves as weak and listless—the idea that they should no longer be "cultivated" to be insipid, useless individuals.[13] Bewell observes that "fundamental to Wollstonecraft's critique of male "gyneculture" is the recognition that plant "luxuriants" and "varietals" rarely occur in nature; instead, they were produced by special horticultural techniques, notably by increasing the richness of the soil or by plant splicing."[14]

The notion of "planting" a young girl in too rich a soil clearly resonates with eighteenth-century concerns about raising young women to be sober and responsible adults. The richness of soil versus richness of society, food, drink, or dress is an easy comparison to make, and one that hints at deeper parallels between womanhood and planthood in eighteenth-century cultural consciousness. The practice of "splicing," at once a complicated scientific procedure and a hotly contested cultural practice, also hints at fears about a woman's bodily fragility and permeability: her weakness to exterior influences ranging from the charming words of a seducer to an excess of claret, and her literal penetrability as a woman. Wollstonecraft is thereby also critiquing the botanist or horticulturalist who would produce such luxuriants, as well as the men and women who would raise young girls in such disastrous ways.

This fear of women becoming like luxuriants does not, however, preclude an association between women and native plants, which were often thought of as possessing any number of positive traits that could easily describe ideal women: hardiness, purity, simplicity, and goodness. George elaborates, "Native flowers were invested with virtue and used for moral teaching in periodical literature for women and in pedagogical texts featuring young female protagonists."[15] Further, to be compared to a native flower—a poppy, a primrose, a cowslip—was to be compared to something essentially British, a series of positive qualities that personified not only an ideal female temperament, but also an ideal national identity. Wildflowers were of particular pedagogical value: they embodied the pure simplicity of something grown without art in nature while playing a role in the perennial division between urban and rural life, between the ultrarich and those with modest income. Unlike the luxuriants and hothouse plants that were staples in the vases of upper-class London households, bouquets of modest wildflowers were more likely to be found in the home of a middle-to-lower-class family outside of the metropole.[16] The fact that native flowers were thought to possess greater pedagogical value for middle-class women also rejects standards of education and deportment as modeled by an increasingly isolated aristocracy. In a period where most educational texts for women and children were written by women, by the late eighteenth century native flowers took on special significance to the female study of botany.

The idea that women could be like plants was thereby not only embraced by men, but also by women who saw the greater potential in such a linkage as authors. From poets to collectors to textbook writers, women utilized this cultural linkage between women and flowers to various positive ends. As I will elaborate throughout this chapter, the relationship between women and plants was inevitably complex and unquestionably fraught. However, it was also a relationship that some women seized upon as useful, taking advantage of the associations between women and plant life to trace larger discussions of wisdom and knowledge, faith and reflection, education and industry. Particularly in botanical textbooks—framed within a larger scientific context that had to account for conventions about knowledge transference and taxonomy—women authors were able to allude to the linkage between women and plants as generative rather than detrimental.

This affiliation between women and plants often transcended the relationship of botanist and subject in these female-authored textbooks. Wakefield and Jacson recognized the cultural allegiances between women and plants as well as the physiological ones, comparing anatomical and biological systems as part of a broader endeavor to illuminate botanical study for their readers (albeit in a far less salacious manner than La Mettrie or the pornographers). Further, as part of what I argue is an explicit recognition of the connections between femininity and vegetality, these authors also explored the critical possibilities of what were known as "monster" plants. Plants condemned as "monstrous," as in the second epigraph to this chapter, could encompass a number of botanical categories: they could refer to plants with doubled or multiple blossoms, plants with abnormal coloration, or plants with ostensibly atypical behaviors such as insectivorous habits or parasitic growth patterns. Rather than serving as botanical subjects in these textbooks, such plants provided evidence of both the ambiguities and curiosities of the natural world by destabilizing taxonomic systems in potentially productive ways for botanizing women. Parsing both the native versus luxuriant divide and the critical potential of those "monster flora," this chapter traces the ways in these textbooks embraced the kinds of vibrant, exciting life that plants could generate for women botanists and readers in the late eighteenth century.

Botanizing Women, Botanizing Authors

Priscilla Wakefield and Maria Jacson were among the most important contributors to the surge of popular women's writing on botany at the end of the eighteenth century. This outpouring of botanical writing by women was actually only part of a larger trend of the involvement of women in the botanical sciences during the Romantic period. As Ann B. Shteir writes, "Encouraged by parents, teachers, and social commentators and pursuing their own interests, Flora's English daughters were botanically active. They read botany books, attended public lectures about plants, corresponded with naturalists, collected native ferns, mosses, and marine plants, drew plants, developed herbaria for further study, and used microscopes."[17] Women were "particularly visible as writers of botany books," with a significant number of botanical textbooks being authored by women, particularly into the nineteenth century.[18] Furthermore, by this time, the seemingly interminable war with France during and after its shocking revolution had resulted in an inward turn to English botany: native plants (as well as efforts at acclimatizing exotic plants) were of greater interest to many botanists throughout the country. This shift in attention arguably provided women with new opportunities to lay claim to scientific authority because of their purported affiliation with such domestic plants, thereby generating new opportunities for authorship and expertise.

The publication of botanical texts by these women is impressive, even given their lifelong interests in science and education and access to expert resources. Jacson, for example, was friendly with Erasmus Darwin through her cousin, Sir Brooke Boothby, and was known for being astute in scientific subjects from childhood. Both Jacson and Wakefield were familiar with the most up-to-date botanical research and publications, citing or referencing botanical volumes published as recently as the same year as their own works. This periodic recourse to male-authored texts as sources of authority that confirmed their own scientific information also had the convenient side effect of deflecting potential criticism of the "learned lady" sort. Kelley observes, "By recognizing the authority of others, [Jacson and other botanical writers] distracted attention from the fact that their own botanical knowledge went well beyond an acceptable minimum."[19] Particularly for the young female readers *Botanical Dialogues* and *An Introduction to Botany* were designed to attract, it was necessary to maintain this balance between demonstrating advanced scientific knowledge and appropriate reserve. Shteir points out that such worries led to the development of very specific models for the botanical textbook and other scientific texts: "Women science writers in the late eighteenth century, like their novelist sisters, wrote for women, used family-centered plots and a domestic locus, and combined narrative forms with didactic intentions."[20]

This focus on family-centric plots and mother/teacher-based dialogues has until recently caused such work to be dismissed as insufficiently scientific in comparison to cognate publications by male authors. While the framing and formatting of such texts did differ from the style of most male-authored publications, the dialogue was a genre with a long history in scientific texts for women, often written by men, including Bernard le Bovier de Fontenelle's *Entretiens sur la pluralité des mondes* (1686) and Francesco Algarotti's *Il newtonianismo per le dame* (1737), both of which were translated into English by women. Epistolary texts were popular in the late eighteenth century as well, and prominent authors such as Jean-Jacques Rousseau had popularized scientific exchanges in his *Lettres élémentaires sur la botanique* (1785). By utilizing a genre with such strong scientific roots, Jacson and Wakefield were actively participating in and contributing to an accepted mode of knowledge transference during the eighteenth century. Moreover, as Mitzi Myers has argued, women's participation in the genre of children's literature (in this case, textbooks written for younger women and girls), rather than delimiting the scope of what they could accomplish, in practice allowed them a freedom and flexibility to address many of the issues considered most important to women in the period in the "new mode of enlightened domesticity."[21] Such flexibility also enabled Jacson and Wakefield to explore different conceptualizations of the relationship between women and plants, both native and exotic. While their professional contemporaries predominantly embraced acclimatized exotics, often at the expense of native specimens, these women made strategic investments in the alliance between British women and British plants.

The debate over the value and significance of native versus exotic plants was a multifaceted one, with botanists and botanizing amateurs weighing in on the subject with some frequency. And, while Jacson and Wakefield engaged in this debate through their creation of pedagogical texts, some authors contributed to the conversation through fictional means. Erasmus Darwin's *The Botanic Garden* (1791) was one such text. Written by the man largely responsible for the Lichfield Botanical Society's immensely popular translation of Linnaeus's *Systema Naturae,* published between 1783 and 1785, *The Botanic Garden* was actually two poems combined into one volume: *The Economy of Vegetation* and *The Loves of the Plants,* and it enjoyed "huge success in the 1790s on both sides of the Atlantic, going into a fourth London edition by the end of the decade, as well as three Irish editions and the American one."[22] *The Loves of the Plants* was the more popular of the two and widely considered both botanically valuable and erotically charged, as evidenced by the outcries among conservative commentators. Despite objections like the ones made by Reverend Richard Polwhele in his *The Unsex'd Females* (1798)—a text that eviscerated women like Wollstonecraft, Barbauld, and Smith, and accused botanizing women of lusting after a "bliss botanic"—this and many other botanical texts expressly designed for women were well regarded among female readers.[23]

Darwin's interest in plants was in no small part motivated by the poetic possibilities of their sexual reproduction. His *Botanic Garden,* particularly the second part, *The Loves of the Plants,* famously illustrated the Linnaean system of reproduction with exuberant fantasies of wanton women left and right, engaged in orgiastic celebrations of plant life and death. In spite of such salacious content, women as readers and as authors still found substantive botanical value in the text.[24] For example, Jacson frequently had both her male and female characters praise the poem as scientifically valuable and easy to access, at the same time as she utilized references to the poem in lieu of explaining delicate subjects in the character of Hortensia.[25] It was simultaneously both a tool and a shield for Jacson in her *Botanical Dialogues,* as it may very well have been for many other women in their personal lives.

Darwin was far from the only prominent author publishing botanical poetry; in fact, many women crafted their own fanciful botanical texts modified by explanatory footnotes featuring taxonomic details and Latin terminology. Charlotte Smith included notes as early as the third edition of *Elegiac Sonnets* (1786), although her investment in scientific terminology is most associated with the unfinished "Beachy Head" (1807), an expansive poem that features detailed endnotes on the plants, birds, shells, and fossils found in the speaker's walks along the shoreline, with a logical focus on exclusively native plants. And, like Jacson, Smith bolstered her own considerable botanical knowledge in these footnotes with citations to male-authored botanical works, such as James Sowerby's *English Botany* and Philip Miller's *Gardeners Dictionary.*[26] In a more distinctive format, Frances Arabella Rowden's *Poetical Introduction to the Study of Botany* (1801) combined prefatory material outlining the essentials of Linnaean botanical study with a series of poems artistically describing three representative plants from each class, utilizing a mixture of native, acclimatized, and exotic plants as subject matter. The extent to which more obscure authors like Rowden contributed their own voices to the broader artistic conversation about a highly technical subject deserves more attention than I can give it here, but it is crucial to note that contemporaries of Jacson and Wakefield were equally invested in botanical study for women and sought to explore the subject through various creative media.

Of course, Darwin's poem was about more than just erotic encounters between anthropomorphized flora, as scholars have usefully demonstrated in recent years. It was also crucially about the botanical endeavors that linked plants (both individual specimens and large groups of samples) to broader discourses of empire, colonialism, trade, and slavery.[27] Exotic plants were increasingly valuable to a British audience for financial as well as aesthetic purposes, as botanical collecting and bioprospecting "both served the colonial enterprise and were, in turn, structured by it."[28] As explorers and naturalists continued to scour the globe for new and rare specimens to bring back to an interested and invested British public, travel narratives, natural history museums and catalogs, and gardening manuals all contributed to the passionate reception of exotic plants.[29] With this in mind, I argue that perhaps the most significant attribute of *The Loves of the Plants* is its breakdown of native versus exotic plants. As Bewell astutely observes, "The nature that appears in Darwin's verse . . . are not common plants, but choice ones. . . . They are as much a product of human invention and industry—as much a novelty—as anything else in the poem. In this sense, *The Botanic Garden* is very much engaged in making nature an integral part of the industrial and consumer revolution."[30] The plants featured in Darwin's signature poem were primarily exotics, imported from other countries both nearby and far away. For example, one of the earliest mentioned exotics is the plant from which the spice turmeric is drawn, curcuma.[31] Native to Southeast Asia and part of the English culinary repertoire since the medieval period, it was also by the eighteenth century part of a trend in herbal remedies for treating women's health issues, particularly dysmenorrhea, or "green sickness."[32] Curcuma was cultivated in many botanical and personal gardens in the East Indies, as it could not successfully be grown in the chill and low humidity of the English climate, even in specialized acclimatization gardens.[33] It was also valuable as a spice or a source of dye and was a factor in establishing trade and/or a colonial relationship with the populations that grew it—clearly an exotic plant of some significant cultural and political value. Other references early in the first Canto include decorative flora such as kleinhovia (a genus of plants found in very narrow regions in Asia), gloriosa (native to Africa and Asia), and, of course, the tulip.[34] Exotic plants are frequently brought to the reader's attention—alongside the most intriguing of European plants such as the rhizomatic marine plant zostera or the "vegetable lamb" barometz—and readers are quick to learn that *The Loves of the Plants* imagines a global garden filled with many such species.[35]

This fascination with exotics, both useful and ornamental, was borne out in botanical trends throughout England. Acclimatization gardens like the Chelsea Physic Garden and botanical projects such as the Royal Botanic Gardens at Kew, as well as the private collections of individuals such as the Duchess of Portland and Lord Bute, all centralized previously foreign plants in decidedly English ways.[36] In fact, as Bewell notes, "the species that made up English nature were fundamentally changed.

Much of this English nature was now coming from somewhere else."[37] Kew was at this point even comprised almost entirely of exotic species. As far as private gardens were concerned, it was considered an achievement to include newly imported specimens in one's collection—and an even greater achievement to successfully cultivate them. There were entire volumes devoted to explaining how to grow exotic plants in English soil, including Philip Miller's frequently updated and republished *Gardeners Dictionary*, which Thomas Martyn (as translator of Rousseau's *Lettres*) enthusiastically recommended in his preface to that same work. Private citizens, too, were more interested in exotics, based upon garden purchases and publishing records.[38]

By the turn of the century, around the time when Wakefield and Jacson were publishing their texts, exotic plants were fully entrenched in English gardens and English culture. Yet these women writers exhibited complex relationships with exotic plant life, particularly with regard to the relationship between femininity and flora. Wakefield in particular expressed a preference for native plants as the subjects of her study, offering the simple and more readily available flora of the English gardens and fields to her female readers. And, although Jacson's stance on exotics versus native plants was more nuanced, she, too, believed in plants grown at the very least in the modest confines of a personal botanical garden. Such a position was likely inspired by practical considerations; it may have been difficult for many middle-class readers of these textbooks to gain access to the private gardens of the aristocracy, or to travel to visit the major public gardens. To highlight only florists' flowers or imported exotics would be to exclude a significant portion of readers from being able to do precisely what these books urged them to do: actively engage in botanical endeavors. Their attention to the fact that women had considerations of both physical and intellectual access, freedom of movement, and financial constraints is also in part what distinguished the works of Wakefield and Jacson (as well as many of their female contemporaries) from that of their male counterparts.

A secondary trend in *The Loves of the Plants* worth exploring in the context of women's botanical writing is its fascination with abnormal or atypical plant life.[39] For example, a striking verse in the first Canto features the "monstrous" plantago or rose plantain, which Darwin fancifully relates to English culture with a reference to *Othello*:[40]

> With strange deformity PLANTAGO treads,
> A Monster-birth! and lifts his hundred heads;

Yet with soft love a gentle belle he charms,
And clasps the beauty in his hundred arms.
So hapless DESDEMONA, fair and young,
Won by OTHELLO's captivating tongue,
Sigh'd o'er each strange and piteous tale, distress'd,
And sunk enamour'd on his sooty breast.[41]

Given that many species of the plantago are native to Britain, it seems a bit odd to link it to Othello, who was himself not English, although he was most recently brought to life by one of the eighteenth century's favorite English playwrights. That said, the descriptors of "deformity" and "Monster-birth" do echo some of the contemporary critical assessments of the Moorish flaws in Othello's character: Charlotte Lennox, for example, in her *Shakespear Illustrated* (1753), calls Othello's "barbarity to Desdemona . . . unnatural."[42] Darwin chooses to highlight the rose plantain in particular, which becomes "monstrous" with its tendency toward double flowers—a frequent reason among botanists to dismiss a plant as "unnatural." The "hundred heads" and "hundred arms," although not literal, hint at this multiplicity of blossoms, like the kind cautioned against in *An Introduction to Botany* (1760) by James Lee, who sneers that monstrous flowers are merely cultivars and therefore not interesting to a botanist (although perhaps to a horticulturist).[43] As Wakefield later elucidated in *An Introduction to Botany*, "monstrous" plants were not a fit subject for botanical consideration.[44] Yet the example of the plantago is just one of many in Darwin's *The Loves of Plants* that might be considered monstrous; it is clearly a topic of some interest to both the writer and his many readers.[45]

As this chapter reveals, there is much to be learned from the English fascination with monstrous plants. For women as authors and as botanizers, monster plants represented more than the threat of multiplicity which Darwin imagined menacing the innocent Desdemona-as-flower/ flower-as-Desdemona. While it is true that Darwin fantasizes about the same kind of blurred distinction between woman and plant that Linnaeus expresses in his journals, a counterinterpretation of this passage allows for the prospect of monstrosity as a generative, vibrant state of being. For female readers of this passage, then, is it possible to identify with the profound and proliferating Othello, rather than the hapless and wilting Desdemona? Given that we know Jacson personally read and then later enthusiastically recommended this poem in *Botanical Dialogues*, it is not impossible to imagine that she might have gotten her interest in monster plants from a passage much like this one. Even Wakefield, whose focus on native and uncultivated plant life seemingly precludes an active interest in monster plants, was invested in envisioning plant life and human life as complementary modes of being. As I will elaborate below, both textbooks inspire us to rethink the relationship between women and monster plants in the eighteenth century.

Accessible Texts, Accessible Plants

Priscilla Wakefield published *An Introduction to Botany* as her first entirely botanical work in 1796 (although she had earlier published in 1794 a work called *Mental Improvement, Or, the Beauties and Wonders of Nature and Art,* which touched on natural history and botany) and then went on to author several more texts on natural history, botany, and entomology, as well as the enduringly popular *Reflections on the Present Condition of the Female Sex* (1798). Published under her own name, *An Introduction to Botany* was known colloquially as "Wakefield's Botany" and was widely read for several years. It also sold rather well: "There were at least eleven British editions of Wakefield's textbook (1796–1841), two French editions (1801–1802), and three American editions (1811–1818)."[46] *An Introduction to Botany* is framed as a series of letters written from a young lady named Felicia to her sister Constance—who is spending the summer with far-away relatives—echoing the epistolary format of Rousseau's *Lettres* as well as many scientific periodicals. Felicia's botanical activities are guided by her amusingly named governess Mrs. Snelgrove, who represents the scientific authority in this text.

Unsurprisingly for a Quaker educator, Priscilla Wakefield's *An Introduction to Botany* frames its content in religious terms. Like Charlotte Lennox in the *Lady's Museum,* Wakefield conceptualizes the study of nature as cognate with the contemplation of the divine. On the first page of her preface she declares, "The design of the following Introduction to Botany, is to cultivate a taste in young persons for the study of nature, which is the most familiar means of introducing suitable ideas of the attributes of the Divine Being, by exemplifying them in the order and harmony of the visible creation."[47] And, while her illumination of the connections between faith and botany are not always prominently visible throughout the text, Wakefield also includes a passage in one of the last letters in the volume that revisits this subject, essentially framing the entire text with the linkages between botany and religion.[48] Felicia, in speaking of the specific shapes of fir and yew leaves which are meant to enable and encourage fertilization, affirms that "this is a remarkable instance of design, and clearly proves that all parts of creation, if properly observed, would furnish us with examples of the wisdom of an infinite wise Creator, who not only formed every thing in the beginning, but has provided, in a wonderful manner, for their preservation and increase."[49]

The anatomy of a given plant, both for Wakefield and her contemporaries, was irrefutable proof of a purposeful and thoughtful creation. This linkage between botanical contemplation and religious contemplation was also in part a means by which to enable female participation in science; by declaring that the study of the natural world was always also a meditation on religious subjects, Wakefield defended what might otherwise be construed as an unladylike predilection for scientific subjects. Indeed, in spite of botany's significantly female following at the end of the eighteenth century, it was never a bad idea for women to be cautious in exposing the extent of their scientific knowledge.

Following Linnaeus, Wakefield also advances the prevalent eighteenth-century notion that the order of plants echoes the order of the universe, as an element of a grand design is made manifest in taxonomic schemes. This evokes once again the linkages between plant and human life, and more important, plant bodies and female bodies. Like plants, female bodies were understood to be in their ideal forms pliant, immobile, and submissive, yet also beautiful, fragrant, and pleasing. Furthermore, because plants could be dissected and categorized, much like humans, they seemed to prove the existence of a deity who ordered the world on rational, anthropocentric terms. Michel Foucault has explained these parallels as generating "models of visibility" that enabled eighteenth-century scientists to conceptualize life on a broader scale.[50] In other words, it was only by seeing plants in relation to humans that botanists were able to incorporate plants into a schema of knowledge. For Wakefield, Jacson, and many other female authors, it was also true that only by seeing plants in relation to God were botanists able to truly understand and appreciate them.

This positive correlation between religious study and botanical study also applied to the academic practices of young people. Wakefield soon shifts focus to the educational aspects of plant study, emphasizing the importance of teaching children (here, especially female children) to think for themselves, to be creative and critical at play, and to engage with complex ideas on age-appropriate terms. The preface continues, "Children are endowed with curiosity and activity, for the purpose of acquiring knowledge. Let us avail ourselves of these natural propensities, and direct them to the pursuit of the most judicious objects: none can be better adapted to instruct, and at the same time amuse, than the beauties of nature, by which they are continually surrounded."[51] Evoking many of the same principles frequently espoused by other writers of children's educational books such as Sarah Trimmer, Maria Edgeworth, and Charlotte Smith, Wakefield encourages the active study of nature as appropriate and beneficial for young readers.

However, Wakefield did not consider *An Introduction to Botany* to be targeted exclusively at younger readers. She references the challenges in making botanical content accessible to female readers, presumably of all ages, after lamenting about the struggles women have faced thus far: "Till of late years, [the study of botany] has been confined to the circle of the learned, which may be attributed to those books that treated of it, being principally written in Latin; a difficulty that deterred many, particularly the female sex, from attempting to obtain knowledge of a science, thus defended, as it were, from their approach."[52] Latin, although certainly a language that many women could and did learn, serves here a delimiting function in the regulation of who had access to what kinds of knowledge. While the Latin binomial nomenclature was widely considered at least reasonably accessible to women and nonelite botanists, it was still true that in the late eighteenth century many of the most important botanical texts were written entirely in Latin, including Linnaeus's own *Species Plantarum* and *Systema Naturae* (which emphasizes the incredible importance of the Lichfield translations). This obscurity was also exacerbated in part by financial barriers; Wakefield notes that much of what was currently on the market was "too expensive, as well as too diffuse and scientific" for young readers, and implicitly, for the mothers who might read such content to teach their children or for their own edification.[53] While the insular and secretive nature imputed to botany could have been an accusation leveled at most of the sciences in the eighteenth century, botany was in actuality one of the few that did eventually achieve a level of transparency in the period through the work of authors like Wakefield and other scientific popularizers. At the outset, her volume was designed to be more accessible to women as teachers and as readers.

Her purposeful abstention from Latin botanical terminology, coupled with a preference for bowdlerized names for plant anatomy, aligns Wakefield's text more closely with works like William Withering's *Systematic Arrangement of British Plants* (1796) than we might expect. Like many other popularizing texts, *An Introduction to Botany* worked to appeal to broad demographics of scientifically inclined women and, in Wakefield's case, to address significant concerns over the pursuit of scientific knowledge. For example, the characters in the text represent real women's anxieties about accessing botanical Latin: as the character Felicia complains to her sister Constance, "I am impatient to make a beginning, but am full of apprehension of the number of hard words at the entrance."[54] As part of her dedication to avoiding "hard words," Wakefield consistently avoids technical terminology, with a few notable exceptions: in rare cases, she slips in references to Latin terminology, as she does when describing the "the dypsacus, vulgarly called teasel."[55] Wakefield also typically utilizes vulgar terminology in-text and then provides a footnote with the Latin terminology, as she does with the plants Madder (*Rubria*), Goosegrass (*Galium*), and Dodder (*Cuscula*).[56] She is, however, inconsistent in her usage of anatomical vocabulary, utilizing more Latinate terms like "calyx" and "corolla" while simultaneously preferring the bowdlerized "pointal" and "chive."[57] In the second letter between Felicia and Constance, the former carefully describes a flower to her sister with deliberately nontechnical vocabulary, using ideas such as supports, tendrils, floral leaves, and scales, instead of their formal correlatives peduncle, receptacle, corolla, and so on.[58] Yet, by the fourth letter, Felicia is deliberately "examining pointals and stamens" in another mixing of registers. Without any real uniformity in terminology, Wakefield's book is possibly at the outset somewhat confusing for readers, although this evidently did not affect its sales.

Felicia continues her correspondence in a third letter that lays out the essential components of a plant, using both the accepted Linnaean terms and the bowdlerized versions made popular in adaptations for women. In this missive, Wakefield demonstrates her own extensive knowledge about botany—a knowledge she needed to write this book and was obligated to deemphasize in order to maintain the appropriately feminine comportment she needed to sell it. The letter includes two charts, the first of which describes the "seven parts" of the plant. Latinate terms are in parenthesis, and vulgar terms are placed outside the parenthesis; the terminal letter refers to the botanical illustration enclosed in the letter.

1. The (calyx) cup, or empalement, *a*.
2. The (corolla) blossom, petals, or flower-leaves, *b*.
3. The (stamina) threads or chives, *c*.
4. The (pistillum) style or pointal, *d*.
5. The (pericarpium) seed vessel, *e*.
6. The seed or fruit, *f*.
7. The (receptaculum) receptacle, or base, *g*.[59]

Here we see that despite Wakefield's prefatory claim to have produced a text that was "divested as much as possible of technical terms," it is clear that certain elements of technical communication were utilized in the production of the text—not least of which were these Latin terms.[60] Henrietta Shirk has made a compelling case for assessing Wakefield's work as an example of highly effective technical communication, arguing that elements of the text such as detailed illustrations, the inclusion of a glossary and tables (like the one above), and a clear writing style all contributed to making the volume more accessible to less highly educated readers.[61] I want to build upon this useful assessment by proposing that Wakefield is also crafting a text that is an astute critical interpretation of an entire body of existing botanical writing by and for men. By distilling the complex elements of Latin terminology into a table with common terms, she both reflects her own knowledge of more complicated botanical information and makes that knowledge a shared repository among female readers.

As part of her production of a technically adept yet accessible text for female readers, Wakefield also included a number of beautifully illustrated black-and-white plates that she likely drew herself.[62] The plate accompanying the above table, for example, features the anatomical elements of a lily—one of the most common botanical subjects presented to women. In this illustration, the full blossom is presented alongside microscopic dissections of the smaller reproductive parts, each correlating to a line in the chart with both Latin and vulgar terminology. The inclusion of plates such as these throughout the text (eleven plates in all, most of them printed in an appendix at the end of the volume) echoes the format and style of many male-authored botanical publications.

Unlike some of those texts, however, which purported to be the sole resource for botanizing women, Wakefield encouraged her female readers to use the plates only in conjunction with actual walks in the fields and gardens and with active experimental endeavors. Felicia explains to her sister, "In order to assist you in the examination of the minute parts of small flowers, it will be necessary to provide a magnifying glass, a needle, lancet, and a pair of small scissors, to render the dissecting them easier . . . a pair of small nippers will be a useful addition."[63] Wakefield's characters and readers are thus not merely absorbing botanical knowledge from an expert source; they are being asked to actively involve themselves in the pursuit of such knowledge and to replicate the kind of experiments done by some of the most accomplished professional botanists. Just as important, however, the recommended tools and technology required of

Constance on her self-guided explorations are items that most women would have had around the house for domestic purposes—for example, needles, scissors, a knife. And although Felicia does recommend the use of a microscope for certain experimental endeavors, such as the examination of tiny daisy seeds, it is not a tool that is emphasized as a necessary part of a woman's botanical arsenal, and at no point does Wakefield recommend that her readers procure one.[64] Once again, Wakefield's text remains more accessible to her female readers than many other contemporary botanical publications.

In content, however, *An Introduction to Botany* remains fairly similar to books like James Lee's *Introduction to Botany* (1794) and Colin Milne's *Indigenous Botany* (1793). The bulk of the volume is devoted to explaining the taxonomic classification of plants according to the Linnaean system with detailed examples of plants found in each class, and by the fourth letter Felicia has moved on to describing some local plants, such as "the Crown Imperial, the Stock Gilliflower, and the Pea."[65] While the Gilliflower and the Pea are native to England, the Crown Imperial (*Fritillaria imperialis*) is a native of the Levant. It had been cultivated in England, however, for at least the latter half of the eighteenth century: it appears in John Edwards's *The British Herbal* in 1770, and William Curtis's *Botanical Magazine* in 1792 and is not described by Felicia as anything new or exciting. As the letters continue, it becomes evident that nearly all of the examples Felicia provides for her sister Constance to study (and those to which Mrs. Snelgrove has directed her attention) are of native or acclimatized plants, from the grasses in class Triandria to the ferns and mushrooms in class Cryptogamia. In one letter, Felicia even explicitly states, "I intend to select our examples from plants of British growth."[66] Throughout, *An Introduction to Botany* continues to avoid those "hard words" while emphasizing the importance of evaluating native flowers.

The study of botany in this text is, therefore, also specifically and crucially a study of native plants, valorized as simple, honest, resilient, and, above all, characteristically British. Like Lady Charlotte Murray's *The British Garden* (1799) or William Curtis's *Flora Londinensis* (1777–98), *An Introduction to Botany* was in essence an introduction to British botany. As is evidenced in Wakefield's text, botany was both a safe and appropriate activity for women to pursue, but it was also, importantly, an activity they could and should pursue at home—in the fields, woods, and gardens of the English countryside. As Felicia declares in her very first letter, the subjects of her study are to be exclusively "the natural specimens crowding in fields and gardens."[67] Wakefield rarely mentions exotics in the entire volume and, when she does, neatly dismisses them: in one example she writes, "Rhubarb, so well known as a stomachic . . . is successfully cultivated in this country; but as its extraction is foreign, it does not come within my plan to give a further account of it."[68] A lengthier passage more thoroughly underscores both the significance of native plants to Wakefield's botanical program and the genuine pleasure the study of native plants brought to her female characters. Felicia writes enthusiastically to her sister:

> How often have we walked through the meadows and pastures, without opening our eyes to the wonders they contain! We were, indeed, delighted with gathering a bouquet of the gayest flowers we could collect, and sometimes admitted a piece of grass, for the beauty of its pendant head. But we little thought, that every single blade of these apparently insignificant plants . . . bears a distinct flower, perfect in all its parts; nay, more complete than the fragrant Lily or the gaudy Tulip, and only requires to be nicely viewed, to excited our value and admiration.[69]

Like the conservative commentators who frowned upon horticultural marvels, Wakefield rejects the "gaudy Tulip" as a showy exotic, imperfect and incomplete. The true joy of botanizing for women, Wakefield claims, is finding *both* beauty and intellectual fulfillment in a simple bouquet of wildflowers and meadow grasses. Felicia inspiringly finds "wonders" in the simple solitude of the British gardens and fields, turning something "apparently insignificant" into something special and incredibly important—not only to her program of botanical study, but to her happiness and well-being. Indeed, taking into consideration Felicia's loneliness at being separated from her beloved sister for an entire summer, it is important to acknowledge that the mental support and emotional strength gained from finding and examining such a plant is priceless to this young woman.[70]

Given what we know about the population of plant life in England at the time and the percentage of exotics that had become part of the British botanical landscape, this emphasis on native plants seems at best partial. Taking into consideration the slippages in the text between Latin and vulgar terms, native and acclimatized plants, however, it is unlikely that Wakefield actually failed at presenting a complete portrait of British botany. It is far more plausible that she found significant value in

English "fields and gardens," a type of value that was almost entirely rejected by elite botanizing men like Joseph Banks or Daniel Solander, who considered the frontiers of botanical labor to be found in exotic locations abroad. For Wakefield and Jacson, the labor of botany was generatively undertaken in local, often female, spaces. The garden, as outlined throughout this volume, had special significance for women botanizers. While formal gardens at grand estates and official botanical gardens were typically the provenance of men (with the exception of women like the Duchess of Portland who had significant personal resources), at-home gardens both functional and decorative were often under the direction of the lady of the house: whether an artfully arranged patch of wildflowers or a kitchen garden, it served as an outlet for women's creativity and industry.[71] It was also a place almost completely appropriate for women to inhabit, both in same-sex groups and in mixed company, actively botanizing or merely enjoying a stroll.

Gardens were thereby the locus for native plants and for women, tied together in their rootedness (in multiple ways) in ideologies of domesticity. Just as the epistolary format of *An Introduction to Botany* ties the entire study of botany to the home and female-oriented spaces, the emphasis on studying this delightful science in the garden reaffirms a woman's freedom to pursue scientific subjects. In a book intended for younger readers who were under the guidance of a more knowledgeable mother figure (in this case, a governess), this layering of female authority becomes even more crucial. Notably, unlike in Jacson's text, there are no men in this book; the characters are all female. Moreover, Wakefield cites or quotes far fewer male authors as sources of authority for her own text, with the exception of Linnaeus, to whom she attributes the organizational scheme being elucidated in the textbook. Wakefield thus puts forth a program of rigorous botanical study directed by and intended for women, an intellectually assertive act that had serious financial and cultural benefits. Women loved and read *An Introduction to Botany* well into the nineteenth century, and Wakefield had a reputation as a pedagogical expert long after her death.

Exploring Botanical Boundaries

Maria Jacson was quite accomplished in the subject of botany from her youth: she not only published on botany, but drew flowers, planted her own garden, and conducted horticultural experiments in that garden. She was also influenced by Erasmus Darwin at a young age and wrote four botanical volumes with his approval—he would later recommend her book specifically in his *A Plan for the Conduct of Female Education* (1797), and she frequently endorsed his work in *Botanical Dialogues*.[72] Jacson did, however, publish anonymously and only put her name to her work by the third edition of *A Florist's Manual* (1816), which was her final (and most successful) published work. *Botanical Dialogues* did not pass beyond the first edition, possibly because it was considered too advanced for children and had not managed to attract the notice of more accomplished botanists, due to its declared audience. She then rewrote the material as *Botanical Lectures by a Lady* for an adult audience, which scarcely sold any better. However, in spite of its commercial failures, *Botanical Dialogues* represents an important contribution to the critical conversation around women and botany through Jacson's innovative depiction of women as educators, botanists, and gardeners.

Among its many strengths, *Botanical Dialogues* impressively exhibits Jacson's intimate understanding of the Linnaean botanical system in its comprehensive treatment of elementary botanical principles. It is framed as a series of dialogues between a mother, Hortensia, and her four children, Harriet, Charles, Henry, and Juliette, and was designed to mimic the types of pedagogical treatises often presented to women and children in the period. Jacson thus echoes Wakefield's *An Introduction to Botany*, a text she had likely read, as she attempts in format (a dialogue between a mother and children) and in style (a relaxed, conversational model of disseminating information) to attract a similar audience, of young people. The advertisement for the text, appended to the front of the volume, explains that Jacson published her work "with the hope that her endeavors to render the science of Botany a more amusing and less difficult study to young people, than it has hitherto been found, may not prove wholly ineffectual."[73] As noted, however, Jacson was less successful in making her content accessible to her target audience.

This lack of commercial success has typically been attributed to the level of difficulty of the text itself, specifically because Jacson's use of Linnaean terminology and concepts exceeded the level of knowledge typically assumed in botanical texts for women and children. An explanation is offered for every technical term listed in the text, but it is often given once and never revisited. There are very few tables or charts, but Jacson does include a number of plates that illustrate various plants and details of plant anatomy, scattered throughout the different dialogues, as well as some appended to the volume. She utilizes botanical Latin with very little recourse to vulgar names, and only then in reference to the common names of plants, and never for plant anatomy or classificatory terminology. As Hortensia cautions both her children and the reader, "I apprize you that you will not find the first part of the study of botany particularly entertaining."[74] And while it was precisely her use of exact Linnaean terms that won her the praise of Darwin and Boothby, it was also likely that such insistence upon Latin terminology caused the disconnect between her intended audience and her actual audience. As mentioned above, the volume ran to just one edition in her lifetime, which must have been both a disappointment and a surprise to her, given her extensive knowledge in a wide variety of biological sciences, from botany to ornithology. The significance of Jacson's dedication to the Linnaean system, however, transcends questions of commercial viability and demonstrates how women could creatively interpret and reinterpret such classificatory systems for their own intellectual and artistic ends.

In utilizing the Linnaean classification and in directing her pupils to do the same, Jacson hoped to invite her audience into the broader system of knowledge acquisition and organization that had shaped the face of the natural world over the course of the eighteenth century. For Hortensia, it was not that Latin was a difficult language for women and children to acquire (her eldest son, Charles, was in the process of learning elementary Latin at school), but it was the hard truth that Latin was quite simply the gatekeeper of botanical knowledge. In order to gain entrance into what she very clearly envisions as a wonderland of knowledge and information, one must learn a little botanical Latin. Hortensia explains to her children, "Pray observe, that I intend strictly to require the use of Linnaean terms, as that will be a means of imprinting on your minds what you learn, and as you grow older, will make you ready in the language of botany."[75] This idea of a "language of botany" underscores the obstacles that women had to overcome in the pursuit of scientific knowledge. Yet, crucially, the language of botany was for Jacson something informed by, but distinct from, Latin. As the table of contents summarizes, "The Linnean terms ought to be made use of; knowledge of Latin has no great use to Botanical pupils. Botany has a language peculiar to itself; an agreeable study."[76] Here, botanical Latin is distinct from Latin in general, and as the botanical science "has a language peculiar to itself," it is a language that women of all ages are quite capable of learning, especially with the help of useful translations like those done by Darwin and his compatriots at the Lichfield Botanical Society. Hortensia even reassures her son Charles that he will not be expected to have an advantage over his sister Harriet, elaborating to all her children that "Charles will know that calyx means cup, but that will not assist him in the various species of calyxes, which he will have to retain in his memory, the common meaning of words is not sufficiently precise for the purpose of science."[77]

Jacson's endorsement of Latinate Linnaean terminology, while strident, is particularly considerate of additional challenges of access faced by women. While Charles does indeed know that "calyx" means "cup," and how to specifically pronounce "calyx," words expressly intended "for the purpose of science" are challenging for both male and female readers. As Jacson makes explicit, the difficulty of many terms arises in part because of the lack of accents on the words, indicating which syllables ought to be emphasized in pronunciation.[78] She again compliments the Lichfield translations for their accents, which "greatly facilitate the use of them," and observes that "those very people, who object most to the difficulty of them, speak without hesitation of convólvulus, geránium, aspáragus, campánula, and many other names, which are all of them the Linnean ones."[79] As evidenced by the preceding quotation, Jacson herself makes a point of using accent marks throughout the entire text—an incredibly important stylistic element that makes an otherwise intimidating set of Latin terms more accessible to disadvantaged female readers. Reassuring both her daughter and her readers that they will appear neither pretentious nor ignorant with a judicious use of Linnaean terminology, Jacson-as-Hortensia reveals the study of botany to be far more approachable than many women likely believed.

The question of course arises: Why does *Botanical Dialogues* not propose the study of anglicized botanical terminology? Such adaptations were much more common throughout the latter part of the eighteenth century and were easy to sell in a marketplace with an unprecedentedly high demand for botanical works. Even the character Harriet remarks to her mother, "Sometimes I feel a little afraid of being found dull; and I think I have heard of botanical books written for ladies, which make all the hard words easy."[80] Jacson answers the question in the voice of Hortensia, noting sharply, "There are some books, which pretend to do it; but the scientific terms are still to be learnt, and when learnt, in the language of those books, you cannot converse with a Linnean botanist; they may make you a partial, but cannot make you a universal botanist."[81] It is worth observing here that although she does not completely dismiss Withering's work (she does recommend his *Arrangement of British Plants* for its treatment of cryptogams), Jacson slights the entire concept of bowdlerized vocabulary by dismissing the notion of "partial" botanical knowledge.[82] In this way, she is also thoroughly dismissive of the abundance of botanical texts written expressly for women, condemning their information as "partial" at the very best. This moreover seems to suggest that Jacson found such targeted texts exclusionary rather than helpful; instead of initiating women into the universal language of botany—with which they might converse with the most learned men in the world—such texts only gave women a "partial" and therefore fundamentally flawed botanical education. Jacson thereby holds her female readers to the same standards as she would hold any male reader, just as Hortensia holds Harriet and Charles.

It is also likely that Jacson's preservation of Linnaean terminology for her textbooks arose from her aforementioned friendship with Erasmus Darwin. A staunch supporter of Linnaeus's work, Darwin was for some years engaged in a bitter dispute with Withering on the subject of botanical terminology. Withering's adaptation of William Hudson's *Flora Anglica* (1762), published as *The Botanical Arrangement of All the Vegetables Naturally Growing in Great Britain* (1776), preferred what Jennifer S. Uglow has aptly termed "prissy" vocabulary and almost entirely omitted the concept of sexual dimorphism from his text, resulting in what one critic called an "unintelligible" work.[83] Darwin's translations of Linnaeus's work, on the contrary, won him praise in the best publications and from the leading botanists of the period. Jacson, then, was both following the lead of her friend and adhering to what was widely recognized as the best practice in botanical publishing.

There is a relative equality between male and female perspectives on botany that is preserved throughout the text of *Botanical Dialogues*, as the boys and girls ask roughly the same number of questions, inquire on similar points, and engage in all of the same "assignments" that Hortensia asks them to pursue after her conversation with them. Hortensia also introduces the sexual system of Linnaeus in a manner designed to privilege neither the male nor the female reproductive organs. She explains, "The stamens are termed males, and the pistils females; these most frequently exist in the same flower, but are sometimes in different flowers, and sometimes even on different plants."[84] When delineating the Linnaean system of classes, she also omits mention of the organizing logic behind such a system: that is, she does not tell her children that the entire taxonomic model privileges the male reproductive part (the stamen) over the female reproductive part (the pistil). She merely notes casually, "This single [organizing] character is arbitrary, and has been taken from various parts of the fructification by different authors; some have chosen the petals, others the fruit; Linneus has made the choice of the stamens, and on their number and situation has founded his classes."[85] And in so nonchalantly mentioning the Linnaean distinction among other taxonomic systems, she is also, importantly, deprivileging Linnaeus. To Jacson, Linnaean taxonomy may be the most effective system, but it is not absolutely the best one. This ambivalence about the value of using the reproductive organs of plants to organize a system of classification also reflects Jacson's subtle resistance to masculine taxonomic schemes.

Jacson also felt emboldened to point out the discrepancies in and failings of the Linnaean sexual system in *Botanical Dialogues*. In an early conversation between Hortensia and her children, she writes cautiously, "I recommend that you follow Linneus . . . till these small defects of his system was removed by the attention of those, who, knowing its great merit, are more desirous to render it perfect than to expose and cavil at the few errors they can find in it."[86] At once deferential and critical, Jacson is pointing to the flaws in the system while deflecting the potential accusation of any intention to correct those flaws herself. She is also criticizing those male authors who would prefer to "expose" errors rather than work to improve an otherwise excellent system simply because of a philosophical disagreement. In many cases, these "caviling" men were those who preferred the taxonomic systems of John Ray, whose much older *Historia Plantarum* (1685–1703) was still in favor in conservative circles because it was not based upon sexual dimorphism. Simultaneously, however, Jacson littered her own writing with references to other male-authored publications, citing, among many others, sources as diverse and wide-ranging as John Gerard's *Herball, or Generall Historie of Plantes* (1597), Nehemiah Grew's *The Anatomy of Plants* (1692), William Curtis's *Flora Londinensis*, and James Sowerby's *English Botany* (1790). Each of these texts were well regarded, although some were more popular than others for a variety of reasons both scientific and financial and in some cases represented a very current knowledge in the botanical sciences.

Botanical Dialogues is filled with such tensions between Jacson's own clearly impressive knowledge and her desire to remain appropriately modest. A poignant passage from the text reminds female readers that "the world have agreed to condemn women to the exercise of their fingers, in preference to that of their heads; and a women rarely does herself credit by coming forward as a literary character."[87] She elaborates, and not without regret,

> At this time information in a woman, beyond a certain degree, distinguishes her above her companions, and like all other distinctions is liable to lead her into a vain display, of what she hopes will gain her admiration. Hence she becomes ridiculous, and brings, what in itself might be a credit, into a disgrace; whereas the disgrace ought to fall only on herself, and not stamp ridicule upon those of better understandings, who extend the advantages of their education to every occupation of life.[88]

The fear of becoming "ridiculous" and bringing "disgrace" upon herself is clearly at the forefront of Jacson's mind, as she concedes the dangers inherent in expressing too great a familiarity with botanical knowledge. However, as Hortensia has been reassuring Harriet all along, she firmly believes it is the woman who exhibits too much knowledge, not the woman who possesses too much knowledge, who is at fault. Hortensia reminds her daughter that both "education of the fingers" and "that of science or language, of the mind" are equally important to the proper development of the female intellect and are "becoming the female character."[89] As attractive as the botanical sciences are, they cannot be privileged over work "of the fingers" like sewing or mending. The emphasis on the fine balance necessary to navigate a society all too ready to dismiss educated women as "unsex'd females" is a crucial component of Jacson's project in *Botanical Dialogues*. Wishing to avoid the fate of outspoken intellectual women like Wollstonecraft, dismissed as a promiscuous plant in a dialogue written by Polwhele and fellow conservative Thomas James Mathias, Jacson puts forth a plan of botanical study for women that achieves an admirable equilibrium between rigorous intellectual activity and showing deference to male authority.[90] At the same time, a closer reading of her work reveals an understated resistance to male intellectual authority in her willingness to critique the entire basis of the sexual system of classification.

Linked to this resistance to male authority is Jacson's philosophical stance on appropriate subjects of study for female botanists—a position at once similar to and more daring than peer publications. While Wakefield includes in her textbook an explicit indictment of exotic or acclimatized plants, Jacson's take on native plants is subtler. She never forbids the study of exotics in the same way that Wakefield seems to, but Hortensia prefers to let her children explore the world around the home, in the fields and her very own botanical garden—perhaps very similar to the one that Jacson herself kept. Hortensia's "flower garden," equipped with "books and glasses," indicates her advanced level of interest in botany and reflects her dedication to preserving that balance between domesticity and scientific activity.[91] Many women (even with few resources) certainly kept a flower garden, but it is the instruments through which Hortensia encounters her garden that informs her botanical approach: with glasses for microscopic viewing, books for consultation, and additional tools for dissection and collection, she is as prepared as any male field botanist. She is likewise content to provide examples for her children that are largely comprised of native plants, from grasses to flowers to fungi. In each of the separate dialogues, she uses reference to native plants to explain the reproductive processes and development of plants, as well as the Linnaean system of classification.

Harriet and Charles, representing the novice perspective in the textbook, often range further afield than their mother's garden, gathering specimens from the meadows and forests while playing with their younger siblings. As Hortensia lectures on the different processes of fructification, for example, they recognize the various stages of plant development by associating the process with a plant they have seen outside both in the wild and in their mother's garden, such as "snap-dragons," "larch," and "fraxinella."[92] The children, unsurprisingly, mix registers in their references to plants, occasionally remembering the scientific names and in other cases lapsing into the vulgar names they had learned before. Hortensia often reminds them to exclusively utilize botanical terminology, but in one of the charmingly authentic aspects of Jacson's pedagogical text, they often forget. They remember, however, their mother's injunction to study botany by doing, rather than exclusively by reading. Like Rousseau, who declared grandly that "botany is not to be learned in the closet; you must go forth into the garden or the fields, and there become familiar with Nature herself," Jacson reminds her readers that hands-on instruction is best.[93] She also notes sagely, "As we do not unfrequently see great and wise men err in their judgment and accounts of things, we must not rely upon them as infallible: in whatever you undertake, make it a rule to see for yourself."[94]

Consequently, Hortensia instructs her children to collect specimens, observe and analyze them, and report their findings back to her before the start of each dialogue. They are even promised the option of creating their own "experiment-garden" should they progress enough in their studies within the next year.[95] After they conduct their elementary experiments, Hortensia then leads them in dissection. For example, in the second dialogue, she directs her eldest son, "As a reward for your great attention, we will dissect a flower; the parts of the crown imperial are so large, that it is well suited to our purpose. Be so good to gather one, Charles, also bring a poppy and a tulip at the same time."[96]

This parental command highlights a key difference between Jacson's model of instruction and Wakefield's. Hortensia's garden, it turns out, in addition to natives of Britain like the poppy, features horticultural darlings like the tulip. And, unlike Wakefield, who so directly dismissed such plants as "imperfect" and therefore unsuitable for botanical study, Jacson is content to use this acclimatized plant (and another acclimatized flower in the crown imperial) as an example for dissecting the calyx of a flower. Lady Charlotte Murray notes in her *British Garden* that the common tulip was first cultivated in England in 1577 by James Gerard (probably referring to John Gerard, of Gerard's *Herball* fame).[97] By the time Jacson was writing, then, the tulip was fully part of the British botanical gardens and British botanical imagination—as well as having been the cause of one of the most dramatic speculative bubbles in European history via the Dutch tulip mania of the early seventeenth century. Tulips were immensely popular horticultural specimens, cultivated in various colors, some of which were variegated. Variegation in plants, while appealing to florists and individual collectors like the Duchess of Beaufort (whose expansive early eighteenth-century herbarium is an excellent example of pre-Linnaean botanical collecting), is actually an abnormality, typically caused by a lack of sufficient phosphorus in the soil or a viral infection.[98] Variegated blossoms are not recognized as a separate species by botanical experts—nor were they in the eighteenth century—as they are considered only variants, but they can be replicated with proper tending and horticultural technique. As Lynch also explains, such blossoms (and their endlessly proliferating varietals) become immensely popular with commercial horticultural endeavors: "Variety, rarity, curiosity, and fancy had become . . . the up-to-date horticulturist's watchwords, thanks both to new methods of plant transfer and to experiments with cross-pollination and forcing."[99] And so, while gardeners prized such variants as evidence of their skill in plant breeding, more conservative critics like Alexander Pope thought of such plants as nothing more than deceptive artificiality. For example, in his "Epistle to a Lady" (1743) Pope sneers, "Ladies, like variegated Tulips, show, / 'Tis to their Changes half their charms we owe."[100] Likewise, Jonathan Swift in "The Lady's Dressing Room" (1732) is horrified by the disingenuousness of female beauty, writing of "gaudy tulips raised from dung."[101] In women and plants, natural charms are much preferable—according to men, of course.

While we do not know if the tulip Hortensia dissects with her children is variegated, it is unlikely to have been so. The boys' tutor, Mr. Wilson, expresses an active interest in botany, but there is no indication that the family employs a professional gardener. And although Hortensia references her own "experiment garden," the types of experiments she mentions are dissection- and preservation-based, rather than actual genetic experiments. Thus, Hortensia's specimens may not all be native plants, yet she still prefers to utilize botanically sound examples in her lessons, rather than aberrant plants with misleading characteristics. Further, in her rejection of the excessive florist flowers as valid botanical subjects, Jacson also continues to resist the detrimental metaphorical associations between women and plants and to fight against what Lynch calls the "compact of both reality and fiction" in cultivated flowers, "at once the stuff of Nature—Nature's gift—and an artifact of human fancy and fetishism."[102]

In this respect, Jacson's advocacy of botanical research aligns more closely with Wakefield's philosophy. Both women believed that there were more limitations to what constituted an appropriate botanical specimen than simply the presence of all subsidiary and reproductive parts (the typical measure of usefulness to professional botanists), but simultaneously found such limits generative, rather than restrictive. By embracing a specific population of plants for botanical study, Jacson encourages selectivity, discernment, and sound judgment—positive qualities for any woman to cultivate. The tulip is determined to be botanically useful to Hortensia, not because of its popularity among horticultural circles or subjective attractiveness, but because of its accessibility to female botanizers. Her openness to non-native species, moreover, expands the scope of women's potential botanical projects—and therefore their intellectual labors—as it simultaneously recuperates the value of certain types of exotic specimens to English botany as a whole project.

Further, unlike Pope and Swift's tulip-as-painted-lady, Jacson's tulip is a botanical specimen and a subject of experimentation. Rendering this singular specimen scientific rather than sexual, Jacson reassigns cultural value to the tulip through her writing: the tulip in this text has real scientific worth to female botanists and to readers of *Botanical Dialogues*. In using the tulip as a subject of scientific experimentation, Jacson also reclaims the linkages between women and flowers: the metaphorical and literal association is not one of sensuality or ornamentation, but now one of active intellectual labor and a genuine pleasure. Offering a counterpoint to Wollstonecraft's fears about cultivated luxuriants, Jacson imagines an easily achieved scenario in which women and tulips could happily coexist, just like women and poppies. By reframing this relationship between women and plants, both native and exotic, Jacson imagines a world in which the already embedded female botanist—always a representative of and a part of nature in a way that male botanists were not—can explore anew the boundaries between human and vegetal being.

The Lives of the Monster Plants

The passage from Wollstonecraft's *A Vindication of the Rights of Woman* with which I began this chapter poignantly evokes the cultural, physical, and emotional connections between women and plants in the eighteenth century. While the debate over native versus exotic plants played a significant role in this relationship, such linkages between women and plants become even more complicated in these textbooks as the female characters themselves sketch an ambiguous line between plant bodies and human bodies. As I have alluded to above, the assumption that plant biology was similar to human biology was a popular approach in the period. Elaine Miller elaborates: "The form of the human body—which is an animal body—seems to inform . . . major approaches to nature up until the twentieth century."[103] This was certainly true of Wakefield's work, and, to a lesser extent, Jacson's. For example, Felicia conceives of the circulation of sap in plants as comparable to the circulation of blood in the human body, as she observes that "the sap-vessels, corresponding to the the [*sic*] blood-vessels in animals," and "the fluids of plants are the sap, analogous to the blood of animals."[104] This was a ubiquitous point of comparison, dating back to the early modern period and William Harvey's discovery of the circulation system in animals, which was said to have inspired the quest to discover a parallel circulatory system of sap in plants.[105] In fact, the idea of a circulatory system in plants was disproven by Stephen Hales in 1727, a Fellow of the Royal Society whose *Vegetable Staticks* was well regarded for its survey of plant transpiration. The myth of circulating sap, however, persisted throughout the century in many popularizing texts, and the idea informed many analogist approaches to describing plant biology like Wakefield's.

Felicia continues her analogy with metaphors that move from the interior of the body to the exterior, from the invisible to the unavoidably visible: "The bark of plants seems to perform the same office to them, as the skin does to animals; it clothes and defends them from injury, inhales the moisture of the air, and extracts, or conveys from the plant the superfluity of moist particles."[106] Plants are quite literally in this formulation flesh and blood creatures (or bark and sap creatures, which to Felicia are the same thing). This vulnerability of the plant also crucially ties it once again to animals; the plant as a permeable being is susceptible to changes in humidity, temperature, and physical circumstances, much like the permeable human body. Simultaneously, permeability is accompanied by the same myth of impenetrability as the human body in eighteenth-century culture: bark/skin is a protective mechanism (which is biologically true) that fundamentally keeps the plant/human separate from the nature that surrounds it (which is physiologically debatable and culturally suspect). What this leaves us with are the cultural constructions with which eighteenth-century science surrounded plants both as a group and as individual specimens. Plants were in theory viewed as fundamentally different from human bodies, at the same time as the slippages between them become increasingly evident in concerns over hybridity. It is therefore unsurprising that Wakefield would use this analogy as a writer of botany; what is remarkable, however, is the way such an analogy resonates with the particular intimacy between women and plants. These passages take on an arguably greater significance because this analogy comes from the pen of a woman writer—a woman who herself is inextricably tied with plants in deeper ways, both culturally and physically.

There were certainly other authors invested in sociobiological analogies between humans and plants, including Linnaeus and Darwin as well as Charlotte Smith. Where Wakefield and Jacson differ from them, however, is in their exploration of the concept within texts designed for female consumption; unlike authors with a general audience or a broad scientific readership, these two women highlighted the connections between plant bodies and human bodies specifically for women. To Wakefield and Jacson, the woman-as-plant concept was a generative idea for their female readers that might enable those readers to move beyond thinking of themselves as Wollstonecraft's detestable hothouse flowers. If women are already aligned with plants in English culture, then why not make that alliance a productive one? For women to be characterized as trees—to use Felicia's example—rather than lilies is to project strength, endurance, vitality, and (through the communal nature of forests that scientists now recognize) solidarity with their companions. Here, Wakefield replaces demeaning floral metaphors with an empowering botanical metaphor. Recall Barbauld's "To a Lady, with Some Painted Flowers," in which trees are masculine, described as "sheltering," "tougher," and supportive of Britain's future navy, and flowers are overtly feminine, being "soft," "gay," and "lovely."[107] Unlike Barbauld, however, Wakefield remaps the division between types of plants onto exclusively female subjects, turning both trees and flowers into types of women.

The connections between plants and animals deepen as Felicia elaborates on the function of protective mechanisms like thorns or stingers: "It is worth remarking, as a farther analogy between plants and animals, that the former frequently lose their thorns, &c. by cultivation, as wild animals are deprived of their ferocity, by living in a domestic state, under the government and protection of man."[108] Here, we again see evidence of Wakefield's preference for wild, uncultivated plants as the subjects for botanical study, as well as a surprising sympathy for domesticated plants. Because thorns are designed expressly to "protect [plants] not only against some kinds of insects, but also against the naked mouths of quadrupeds," a removal of those thorns renders plants defenseless, helpless, and dependent.[109] And so, without those thorns, presumably an integral part of the divine plan that Wakefield had earlier identified as the organizing principle behind all botanical life, the plant has become something other than its true self, something fundamentally different (and not to mention botanically useless). It is also conceivable that Wakefield is tacitly advancing a more ambitious case for abandoning cultivation of certain plants altogether by implying that the "government and protection of man" is not always a good thing.

Such explorations of the parallels between humanity and vegetality are coupled in these two textbooks with an uneasiness about abnormal botanical specimens, plants that were often referred to as "monsters." In spite of their mixed levels of interest in using commonly cultivated flowers in their botanical studies, both Wakefield and Jacson were cognizant of the problems that arose in studying cultivated flowers botanically. As Felicia explains to her sister, "Flowers growing wild, without culture, are the most suitable for examination, because those that are domesticated in the rich soil of our gardens, are frequently transformed into something very different from what nature made them."[110] Horticultural techniques, she acknowledges, fundamentally change the appearance of a plant through things like "change of nourishment," different types of soil, and so on.[111] The most basic of changes to appearance was in the size and number of blossoms; most commonly, flowers could be bred to produce double flowers or appear in different, often variegated colors, as explained above. However, the arrangement and number of the reproductive parts of these doubled or enlarged plants could also potentially be changed, making them essentially botanically unrecognizable. Jacson takes particular interest in the idea of the double flowers, calling them "unnatural . . . monstrous forms," and the character Hortensia explains to her children that "when a flower is completely double, it is deprived of its stamens, which commonly expand into petals."[112] Without the right number of reproductive parts, a plant is easily misidentified in the Linnaean sexual system. By doubling the blossoms and the beauty of the plant, its botanical value is erased according to prevailing taxonomic principles—but has it become useless to women like Jacson or Hortensia, who were outside of taxonomic authority? While Hortensia does instruct her children to disregard such blossoms for the purposes of classification, "monstrous" plants become a means of self-expression for Jacson in a way that was not echoed in many male-authored botanical textbooks.

In his widely acclaimed *Introduction to Botany,* James Lee exclaims against the proliferation of horticulturally modified plants. As quoted in the second epigraph to this chapter, he writes that "MONSTROUS Flowers, such as the *Multiplicate, Full,* or *Proliferous,* derive their Origin from natural ones, and therefore are to be considered only as a Variety from *Luxuriance.*"[113] I want to think more critically about the use of the term "monster," both in Jacson's work and in the larger critical conversation about horticulturally derived botanical abnormalities. If it is possible for monsters to exist not only in the animal kingdom or in our imagination but also in the botanical realms, then the idea of monstrosity is actually a linkage between these spaces. Monstrosity is something that allies plants and animals (and, consequently, humans), not something that divides them. As T. S. Miller elucidates the connections between monstrous beings and what such connections mean for those who decide what is monstrous in his analysis of science fiction, "Like their animal counterparts, vegetable monsters in fiction represent a disruption that also works to contain itself, reflecting unease with a hierarchy that places plants at its bottom, even as these same narratives reinforce hierarchy by positioning the monster as evil, aberrant, an error to be corrected."[114] And, while Miller writes of fiction, the anxieties about hierarchy clearly translate to scientific texts as well: Lee, Jacson, and Wakefield are very anxious for their readers to know that double flowers are not good botanical specimens. They are fundamentally "something very different from what nature made them," something abnormal and occasionally very wrong. Simultaneously, the need to so quickly dismiss monster plants as botanical subjects reflects a concern about and a fascination with the idea of botanical monstrosity as a category of vegetal being.

None of these texts make any claim for plants being placed at anything other than the lower end of the hierarchy. Both Jacson and Wakefield heavily emphasize the use value of plants, and the idea that the botanical study of plants should be linked to learning what purpose they might serve in human industry and economy. Jacson in particular spends a great deal of time explaining how less advanced cultures around the world utilized certain plants, such as lichen, mosses, and grasses and regularly reminds her readers of the supposedly serendipitous relationship of humans with nature.[115] She also crucially establishes a hierarchy between plants and animals in these passages—rendering plants valuable not only because of their worth to humans, but because of the way that animals find value in them as well. In speaking of cryptogams, she notes that "when we do not find a particular vegetable useful to the human species, we are apt to regard it in too insignificant a light, so by many are esteemed most of the funguses, whereas they afford sustenance to a numerous tribe of the animal creation."[116] Here, plants are at the bottom of the organic hierarchy, yet they are valuable to Jacson not merely for their curiosity (which has been established throughout the text), but for their usefulness to other species.

Of course, both humans and animals are useful to plants, too. Humans cultivate plants, protect them from herbivorous insects, nourish them, and spread their seeds around the world, to places those plants could have never reached on their own (much to the chagrin of environmental scientists who are now dealing with invasive species). Creatures like bees, hummingbirds, and butterflies pollinate plants as part of their feeding process, as do many other species through accidental contact (e.g., foxes, deer). Yet the hierarchy that places humans above all other life forms remained intact in the eighteenth century, as it largely does today. However, I want to complicate this idea of hierarchy with T. S. Miller's idea about the monster in the place of abjection. While Miller is talking primarily about fictional plants that eat humans like the carnivorous plants in Arthur Conan Doyle's "The American's Tale" (1879), or real plants like the Venus flytrap, his vision of monstrosity in plants can easily be applied to eighteenth-century notions of botanical monstrosity.

Plants were unquestionably below humans in eighteenth-century ideas about consciousness/awareness, intelligence, and motility, yet the genuine discomfort with botanical monsters underscores the instability in this model.[117] Plants were not always easily categorizable or controlled, as monstrous double flowers without "regular" reproductive parts demonstrate. When such plants could not be made to fit into a system of knowledge, the fragility and ultimately arbitrary nature of taxonomy becomes clear. Jeffrey Jerome Cohen emphasizes this instability in relation to animal monsters: "[Monsters] are disturbing hybrids whose externally incoherent bodies resist attempts to include them in any systematic structuration."[118] Dawn Keetley elaborates on such anxieties around plant monsters in particular, noting that "the horror of plants [derives from] both . . . their absolute strangeness *and* . . . their uncanny likeness."[119] Monsters, both real and imagined, threaten the security and stability of taxonomic order, a knowledge system that was, after all, the basis for an entire field of science in early modern knowledge schemes. I want to argue, however, that a discussion of monster plants accomplishes far more than merely raising concerns about their viability as taxonomic exemplars or botanical specimens. Instead, it opens space for the exploration of radical new ideas about plant identity and the relationships that humans might have with plants.

This possibility of revisiting the kinds of relationships that human beings might form with vegetal beings is explored in Jacson's work, as she reimagines empathetic and emotional connections between people and plants. Through a discussion of parasitic plants between Hortensia and her children, *Botanical Dialogues* depicts a scenario in which plants are uncomfortably close to humans. In this passage, Jacson compares parasitic plants (those that depend upon other plants for their sustenance) to "vicious human creatures," elaborating, "I am sorry to find an instance of so much ingratitude in the vegetable kingdom, for the sequel of the history is, that after it has been afforded support and nourishment by a stranger plant, it overpowers and smothers its protector: in this resembling those vicious human creatures who, being too idle to work for their own support, bring their parents to poverty and death, by the efforts their tenderness induces them to make for their subsistance [*sic*]."[120] Parasitic plants are just as morally culpable as those "vicious human creatures" who work their poor parents to death by their refusal to support themselves—a vignette straight out of a sentimental novel.[121] Jacson also, crucially, envisions a kind of equivalency between plants and humans; in this text, "ingratitude" is a quality shared by plants and humans alike. Plants are not inert in this formulation, either; they actively "overpower" and "smother" the plant upon which they depend for nourishment. The analogy also creates conventional familial structures for plants: the "stranger plant" is transformed into "parents" as the plant's wild growth becomes transformed into a purposeful action. By imagining not only a family but emotions for plants, Jacson assigns them a kind of agency we do not typically ascribe to plants: they can make decisions (the wrong ones), and those decisions have social and emotional consequences. Plant growth, of course, often carries with it such consequences. Plants can grow too riotously and take over a tidy backyard garden, or they can grow poorly and fail to provide food. The difference here is that these consequences have repercussions not in the human kingdom, but in the plant kingdom. Jacson is essentially envisioning a world in which plants have a society governed by rules and conventions, by ethics and morals, just like the human world.

This was, crucially, not the majority opinion in the eighteenth century. According to authors as diverse as Royal Society scientists, philosophers, and amateur commentators, plants were simply inferior to humans. Although *L'Homme Plante* is a satire, it provides a useful metric for what eighteenth-century thinkers thought of the possibility of a vegetative soul. In this brief text, La Mettrie specifically points to the ways in which plants were most decidedly different from humans. He exclaims, "What a difference there is between the plants of our species and those which cover the earth's surface! They rival the stars and form the bright patchwork of the meadows, but they experience neither pain nor pleasure. . . . They die as they have lived, without feeling it."[122] Eighteenth-century botanists knew that plants could react to exterior stimuli (being aware, for example, of the mimosa, which curls up after being touched), but did not typically hold the belief that plants were capable of higher levels of sensation beyond what was commonly called "irritability."[123]

Scientists now know this to be untrue. Plants do in fact feel pain; they give off a specific volatile profile (scent) that indicates danger, pain, or other negative sensations in an effort to warn nearby plants.[124] More intriguingly, these volatile profiles are as unique as fingerprints, which deeply problematizes botanical taxonomy. If individual plants are valuable to botanists precisely because of their ability to represent the perfect specimen after which all other plants were modeled, then the plant itself is hardly individuated. It is merely one out of many, chosen because it fits with a platonic notion of that particular plant. Yet it is precisely this notion of an exemplary specimen that the monster plant destabilizes. If plants are not only capable of the kind of abnormality that can result in a monstrous growth, but of a kind of behavioral differentiation that allows a single plant to be distinguished from its peers on a psychological plane, then how can any single plant be said to be a representative specimen? This, among many reasons, was a cause for eighteenth-century anxieties about monster plants and the threat they posed to systems of knowledge organization. For women, always outside of the structures of botanical knowledge production, this destabilization of taxonomic authority may have very well been an opportunity to probe the boundaries of what might constitute "plant life." Jacson's exploration of notions of monstrosity in the plant kingdom is thereby more significant than we might at first imagine.

Jacson's interest in the intellectual and sensory capacities of plants was also inspired by Erasmus Darwin's *Phytologia* (1800), a sweeping study of plant life and behavior. Among many other sections on agriculture, gardening, and horticultural techniques, Darwin incorporates a section on "The Muscles, Nerves, and Brain of Vegetables." In this brief section he outlines the argument for plants possessing musculature—evidenced by their closing of petals and leaves—and a nervous system, which was requisite in order for plants to register the sensations of light, temperature, moisture, or danger that would trigger such muscular movement.[125] Drawing upon experiments involving the application of acid, cutting of leaves, or other irritants, Darwin argues that plants are capable of feeling "pain or pleasure," and posits the question "Does this not prove, that there is a brain of common sensorium, where the nerves communicate . . . ?"[126] He also asserts the possibility of plants possessing "a sensibility to the passion of reproduction," a suggestion that will not surprise readers of *The Loves of the Plants,* but that is also here applied as proof of plant intelligence. If, as Darwin argues, plants are capable of feeling sexual attraction, experiencing pain and pleasure, and enacting voluntary movement, Jacson's attribution of intentionality to plants is a logical next step.

In addition to their uncomfortably human qualities, monster plants like the parasitic specimen Jacson evokes were fascinating to eighteenth-century botanists because of their ability to grow wildly, excessively, indefinitely. Michael Marder identifies this wild growth as a primary reason behind the conventional separation between plant and human life, because humans have typically perceived "[a plant's] outward growth and augmentation, devoid of any final accomplishment, [as] the 'bad infinity' of vegetal temporality."[127] Plants grow in uncontrolled, unstoppable ways, and their growth is practically endless: it is a biological fact that most plants will continue to expand and expand as long as conditions (soil, light, water) allow (hence the reason that gardening is in large part pruning, trimming, and cutting). Marder's "bad infinity" highlights the almost uncomfortable truth about plants, which is that, unlike human bodies, they lack a discrete boundary. While humans have no correlative possibility for "infinite" expansion, many other points of comparison between plant life and human life make this concept an unsettling one: plants are at once so alien and so familiar that our relationship to them is always fraught. Eighteenth-century tendencies to describe plants in anthropomorphic terms can therefore be seen as a tactic for countering this discomfort, this "bad infinity." If the lack of a "final accomplishment" or purpose is one of the most basic characteristics separating plants from humans—in that people theoretically do everything purposefully, whereas plants have no ability to act with intention—then it seems especially important that Jacson's formulation reverses the exact qualities that separate plants from humans. Plants, in *Botanical Dialogues*, can act with intention.

Thus, we might usefully complicate Marder's assessment of historical perceptions by pointing precisely to this passage of Jacson's, in which "vegetable temporality" is—at least for a time—linked to human temporality. If plants can have a type of agency, enough agency to be vicious or tender, then they exist on a temporal plane in which viciousness and tenderness have meaning, in which actions have those all-important emotional and social consequences. The parasitic plant ultimately kills the other plant, and Hortensia and her children, at least, are saddened; perhaps the reader is saddened, too, inspired by that allusive link to the sentimental novels they might also be reading. Plant culture is linked to human culture, and, for a time, plants are imbued with just as much moral culpability for that "bad infinity" as humans might be. Such infinity, after all, is what kills the host plant. Excess of any kind is notorious in eighteenth-century culture for its disastrous consequences; it seems that the natural world of plants, in Jacson's view, is to be held equally accountable for its excesses.

Whether they were "vicious creatures" or merely creatures whose structures uncomfortably resonated with the structures of the human body, botanical monsters were "a feast" for curious readers. To return in closing to T. S. Miller, I want to quote a passage that summarizes the value of the monster plant: "Forcing us to contemplate and confront the vegetal in the human, the monster plant inspires fear as well as offering ways forward for the posthumanist project: the overturning of hierarchies that the monster plant can effect strikes at the root of humanity's instrumentalist domination of plants, because being forced to recognize kinship with plants will inevitably alter how we think about our use of them."[128] Just like Jacson's readers, whose uncomfortable experience in confronting the human in the vegetal ultimately forces them to question the relationship between plants and humans, contemporary readers of eighteenth-century botanical textbooks are obliged to consider the "vegetal in the human." Perhaps instead of following the order of Felicia's analogy—to say that plant fluids are to blood, as bark is to skin—it may be possible to think of the reverse: that skin is to bark as blood is to sap. Keetley proposes that "plants suggest alternative ways of being that challenge the inevitability of (human) being."[129] Building upon this assertion, I suggest that, for readers of these eighteenth-century textbooks, the vignette of the monster plant represents an entire collective population of plant life, all of which can be seen as fundamentally alien and strange when viewed from outside the limited and limiting framework of early modern taxonomy. In spite of the best professional efforts to the contrary, the legibility of plants in the eighteenth century was thereby disruptively ambiguous: plants were at once entirely comprehensible within a taxonomic system and resistant to being completely absorbed into such a system.

In many ways, the textbooks I have discussed attempted to make the ambiguity of plants more legible, while also simultaneously utilizing that ambiguity in productive ways. Jacson and Wakefield actively encouraged their readers to rethink the relationships they might have with plants in texts whose articulation of radical dynamics between women and plant life revealed opportunities for thinking about vegetality in new and different ways. By encouraging metaphoric alliances with native, domestic plants rather than useless luxuriants, these authors allowed women to continue to claim community with plants while avoiding the pitfalls of the stereotypical floral metaphors that permeated eighteenth-century English culture. Furthermore, by outlining the critical possibilities of monster plants and the strange but fruitful ways in which humanity and vegetality are linked, Jacson and Wakefield open up space for new articulations of plant-human dynamics. Thus, Felicia's sap-to-blood analogy, while scientifically inaccurate, crucially reminds readers of the many concrete and intangible relationships that people have with plants. Plants touch every area of our lives, from birth to death, and plant life will inevitably outlast and outlive human life.

Plant life, after all, is far older than human life, and, as Jeffrey Nealon reminds us, "it is probably worth recalling that the biomass of plant life on Earth's terra firma does remain approximately one thousand times greater than the combined zoomass of all humans and other animals."[130] By recognizing a kind of "kinship with plants," as Felicia and Hortensia have done all along, we are able to arrive at a better understanding of the relationship between women and plants in the eighteenth century. There was so much more to the association between women and plants than simply the idea of physical beauty and passivity. In texts like *An Introduction to Botany* and *Botanical Dialogues,* women and plants become linked in the pursuit of intellectual inquiry, emotional healing, and personal fulfillment. In reading the works of Jacson and Wakefield carefully, it seems clear that these authors very much believed that women could be like the good kinds of plants that Wollstonecraft could not imagine: simple, hardy, useful, and, yes, a bit lovely. And while women could be monsters, too, it was far more likely in the late eighteenth century to find botanizing women, intent on classification, categorization, and experimentation in the garden thanks in part to these two authors.

This reminds us botany is at once a means to an end and an end in itself. It is also, crucially, something that builds community. Like the exploratory women in Haywood's *Female Spectator,* women in these textbooks filter their scientific inquiry through their interactions with other women, just as Wakefield's and Jacson's readers are engaging with a virtual botanic community as they study these texts. In fact, as readers or writers, botanizers or collectors, women in the eighteenth century found in such textbooks like *An Introduction to Botany* and *Botanical Dialogues* that most invaluable resource of shared experience and companionship. Such priceless companionship between women as a means of enlightenment and emotional support was modeled in a wide selection of texts (fictional and otherwise) and in real life, as the next chapter on Maria Edgeworth's varied scientific projects will show. In reviewing a selection of didactic texts and fictional productions, chapter 3 underscores another woman's distinctive strategy for participation in and influence on eighteenth-century naturalist discourses—a strategy that emphatically argues for the significance of female-centric communities to naturalist discourses and networks of exchange.

3

Pedagogies and Possibilities

Maria Edgeworth's Scientific Education for Women

> I do not desire to make my daughter a musician, a painter, or a poetess; I do not desire to make her a botanist, a mathematician, or a chemist; but I wish to give her the habit of industry and attention, the love of knowledge and the power of reasoning: these will enable her to attain excellence in any pursuit of science or of literature.
>
> —Maria Edgeworth, *Letters for Literary Ladies*

As the previous two chapters have demonstrated, the opportunity for women to intervene in naturalist developments was rooted in pedagogical discourses and didactic texts, including periodicals and textbooks, and works of fiction, that served as available venues for engaging with scientific knowledge. Maria Edgeworth's varied literary career encompasses a number of explicitly pedagogical texts, several of which engage directly with chemistry and/or natural history, but in this chapter I look specifically at a combination of her didactic writing and works of fiction to nuance her ambitious goal of promoting science as a salutary field of study for women of all ages. The history of Edgeworth's investment in women's education is certainly a fraught one (with class and race playing central roles in determining who might be eligible for such enlightenment), but within the particular scope of her intended audience she advanced a critical agenda that made room for stimulating scientific inquiry within the inexorable patriarchal structures of eighteenth-century British society. While she initially roots her advocacy of natural history and botany in their ability to cultivate women as moral wives and mothers, she expands the significance of naturalist subjects in later works of fiction—including the novel *Belinda* (1801)—to encompass a wide breadth of emotional and social benefits for women. Specifically, Edgeworth uses botanical knowledge and horticultural discourses in her own life and in her fiction to solve problems, cultivate friendships, and solidify familial bonds.

The assessment of Edgeworth's pedagogical objectives as fairly conservative can be attributed in part to her early writing, including her very first published work, *Letters for Literary Ladies* (1795), excerpted above. Yet, when she made this modest push for women's education, she was speaking not as a woman advocating for a rational education for her peers, but in the voice of a man. The character is that of an anonymous gentleman, responding to the critiques of women's education as advanced by a friend, whose misogynistic advice "upon the birth of a daughter" typifies the criticism aimed at learned ladies in the eighteenth century.[1] The friend's antagonistic attitude effectively encompasses all of the usual objections to an advanced women's education in a forty-three-page tirade: they would become too powerful (in obscurely defined ways); they would never find husbands; should they find husbands, they would be terrible wives; they would offend or shock in polite company; and their physical health and beauty would be detrimentally affected. The friend even goes so far as to accuse the new father of the worst imaginable thing: being "a champion for the rights of woman, and . . . the equality of the sexes."[2]

The rebuttal to this long diatribe against women's education, found in the epigraph above, is perhaps surprisingly restrained. Where the original letter-writer objected strongly to women's professionalization, the respondent (the character we are intended to agree with) soothes his friend with a measured rejoinder in "Answer to the Preceding Letter": "I do not desire to make my daughter a musician, a painter, or a poetess; I do not desire to make her a botanist, a mathematician, or a chemist."[3] Yet this disclaimer seems to be contradicted by the very next sentence, which suggests that the skills he does want his daughter to cultivate will "enable her to attain excellence in any pursuit of science or of literature."[4] As curious as this may seem to a modern reader, *Letters for Literary Ladies* is hardly an exceptional case: the tension between an active endorsement of intellectual activity and a hesitancy to promote extensive learning characterizes much of the writing about women and science in the period. What is exceptional about this text, however, is the extent to which Edgeworth utilizes a seemingly conservative position to advance a genuine plan for women's scientific education in a period where such a pursuit was far from universally accepted.

Mona Narain sees *Letters for Literary Ladies* as a text that enabled Edgeworth to "[circumvent] dominant, patriarchal literary authority," establish her own voice in a literary field controlled largely by men, and claim that "women possessed intellectual abilities equal to men."[5] As Edgeworth's first publication and eventually one of her most well-regarded works, *Letters for Literary Ladies* is an important precursor to her later ideas about women's education. The fact that she revised the "Answer" in the second edition (published in 1798), specifically in order to "assert more strongly the female right to literature," further suggests that she considered this text a key contribution to the argument for women's education.[6] I want to expand upon Narain's argument to propose that Edgeworth deploys these male characters to negotiate her early ideas about scientific education for women—ideas that illuminate the extent to which her body of writing is concerned with promoting women's access to scientific knowledge in spite of male interference. In effect, while Edgeworth seemingly presents a restrained attitude toward women's education that diverges from more assertive calls for women's education like those of Mary Wollstonecraft or Mary Hays, her advocacy for active scientific inquiry and "the pursuit of excellence" in the sciences as elucidated in her other writings is in fact quite enthusiastic. Using *Letters* and some of her other early pedagogical writing as points of departure, this chapter assesses scientific education for women as depicted in Edgeworth's *Moral Tales* (1801) and her popular early novel *Belinda*. More specifically, I trace the ways in which *Belinda* promotes a model of scientific inquiry that is dependent upon networks among its female characters in which the women exchange knowledge, resources, and emotional support. I also argue that Edgeworth's use of the novel form as the medium for promoting her philosophy of scientific pedagogy invites imaginative forms of scientific inquiry that expand the possibilities for female naturalists. Finally, I provide a much-needed interrogation of Edgeworth's ambitious scientific curriculum for women—an underappreciated aspect of her pedagogy.

Finding Paths to Women's Scientific Education

Edgeworth's interest and investment in scientific study for women was influenced by the intellectual and commercial climate of the late eighteenth century. By this time, women's access to scientific material was both unprecedentedly expansive and decidedly limited in very particular ways. A surge in the flexibility and profitability of the printing press had led to the publication of dozens upon dozens of educational resources designed for women, including many works explicitly concerned with science (such as the textbooks discussed in the previous chapter).[7] As was the case earlier in the eighteenth century, however, limitations were dictated by a variety of factors including socioeconomic status, level of education, literacy, and access to resources—access contingent upon location, finances, and a family's attitude toward women's education. It is also fair to say that the push for women's scientific education became increasingly conservative, as the linkages between any kind of intellectual pursuit and domesticity were more markedly emphasized.[8] Just as Eliza Haywood and Charlotte Lennox generatively tied the study of natural history to the domestic garden, and Maria Jacson and Priscilla Wakefield endorsed botany in part because it did not take women away from home and hearth, late eighteenth-century writers like Edgeworth supported scientific pursuits that could be accomplished primarily in the home, such as natural history, botany, and chemistry.

Seemingly eternal caveats about women and learning, expressed with chagrin by fictional characters, eidolons, and authors throughout this study, were still included in practically every piece of writing that advocated for women's education. For example, in *Practical Education* (1798), cowritten by Maria Edgeworth with her father, Richard Lovell Edgeworth, the study of science is initially depicted as something to be encouraged because it prevents women from engaging in more dangerous pursuits. The book declares, "Women, who cultivate their reasoning powers, and who acquire tastes for science and literature, find sufficient variety in life, and do not require the *stimulus* of dissipation, or of romance."[9] While noticeably opposing scientific study to "dissipation" or idleness, the hierarchy it constructs between different types of reading is also key. In this formulation, typically "feminine" forms of reading such as romance are rejected in favor of what the Edgeworths promote as new, better types of reading for women, including educational literature and scientific writing. And although several scholars have argued for the value of romance reading in the eighteenth century as a pedagogical tactic for women, many eighteenth-century thinkers would have disagreed, including the Edgeworths. *Practical Education* both endorses and excuses science because it removes women from the path of literary and moral temptation and prevents them from pursuing inappropriate reading practices.

The problem of "bad reading" was an enduring concern for Maria Edgeworth. In many of her works, characters and figures read badly to nearly disastrous results. To name but a few examples: Lady Delacour's "methodistical" reading and Virginia St. Pierre's romance reading (both found in *Belinda*); Angelica's sentimental novel reading in her *Moral Tale* of the same name; and Julia's fashionable, haphazard reading as imagined in the third section of *Letters for Literary Ladies*. Reading badly, however, is not always the fault of the female reader, but sometimes the responsibility of profiteering publishers who produce anything they think will sell—especially with regard to women's education. In *Practical Education*, Edgeworth relays an expansive critique of the current state of female education:

> Confused notions from fashionable publications, from periodical papers, and comedies, have made their way into common conversation, and thence have assumed an appearance of authority, and have been extremely disadvantageous to female education. Sentiment and ridicule have conspired to represent reason, knowledge, and science, as unsuitable or dangerous to women; yet at the same time wit, and superficial acquirements in literature, have been the object of admiration in society; so that this dangerous inference has been drawn almost without our perceiving its fallacy, that superficial knowledge is more desirable in women than accurate knowledge.[10]

Here we see the wrong sorts of "fashionable publications" and "periodical papers" condemned for their irresponsible conveyance of sentimental, rather than practical, knowledge. The right kinds of publications (including, presumably, Edgeworth's own) would not "represent reason, knowledge, and science as . . . dangerous to women," but would understand that such things are appropriate in moderation and within the right context. The implication of such a passage is that women's scientific education cannot or should not be pursued alone, but rather requires the guidance of a discerning parent, friend, or husband. Hester Chapone, for one, agreed with Edgeworth's seemingly conservative permission for scientific study, as she advised a young wife, "If you can prevail upon [your husband] . . . to teach you a language or a science, you will then find amusement for every hour; and nothing is more endearing than such communications. The improvements and accomplishments you gain from him will be doubly valuable in his esteem; and certainly you can never acquire them so agreeably as from his lips."[11] Female education, both Edgeworth and Chapone argue, necessitates a delicate balance between pursuing beneficial subjects of inquiry and doing so to an appropriate extent.

However, Edgeworth's qualifying opposition of "superficial" versus "accurate" knowledge highlights the extent to which her vision for ideal women's education relies upon a thorough foundation in the sciences (among other useful subjects). She elaborates further in the same chapter: "A girl who runs through a course of natural history, hears something about chemistry, has been taught something of botany, and who knows but just enough of these to make her fancy she is well informed, is in a miserable situation, in danger of becoming ridiculous, and insupportably tiresome to men of sense and science."[12] Here, it is incomplete knowledge that is at fault, once again the responsibility of a fashionable education, in which the poor girl has been only superficially introduced to scientific subjects without any real grounding in the topic. Perhaps Edgeworth is critical of publications like the *Ladies Diary* (1704–1840), where scientific information is presented piecemeal and, when read without the guidance of a wiser person, might be detrimental to an impressionable woman's education. In fact, the *Ladies Diary* was a hybrid of almanac, math problems, philosophical questions, puzzles, and even, in one 1706 issue, "instructions on how to make a model of the eclipse of the sun"—hardly the kind of organized presentation of scientific material that someone like Edgeworth with an interest in programmatic education would have preferred.[13] A reader of such a publication might just about exactly fit the description of one who "knows but just enough [about scientific subjects] to make her fancy she is well informed." This is not to dismiss the very real scientific benefits conveyed by this long-running and well-respected periodical, but to point to the extent to which Edgeworth's scientific curriculum relied upon repeated lessons, frequent and wide-ranging reading, and substantive—as opposed to trivial—content.

There are also echoes of Jacson's Hortensia in this passage, where ridicule and "ridiculous[ness]" are among the chief fears of educated women—fears generated by a very real pressure that an outspoken figure like Hays unfortunately experienced for much of her life. Edgeworth clearly sympathized with and respected this apprehension about acquiring scientific knowledge in a period where men like the antagonistic interlocutor of *Letters for Literary Ladies* represented the dominant viewpoint on the subject. However, she is far from direly pessimistic on the subject: in this passage, she also hints at the vital possibility of communicating scientific ideas to experts who might respect and share a woman's intellectual passion. Just as Hortensia endorses botanical Latin for its ability to allow women to converse with "real" botanists, Edgeworth implies that women can hold their own with "men of sense and science" if their scientific education is properly achieved.

The fear of being "insupportably tiresome" is another consideration, one that influences our understanding of Edgeworth's investment in women's education. Where the conventional formulation might have women concern themselves with not being "tiresome" to society (including both men and women), Edgeworth here singles out "men of sense and science" as those standing in judgment of badly educated women. If the primary aim of women is to marry and achieve an ideal domestic state, then obviously women should care most particularly about the opinion of men. While such a perspective seems to contradict the general atmosphere of protofeminist thought circulating at the turn of the century (including authors such as Wollstonecraft, Hays, and Mary Robinson), it is consistent with other work in Edgeworth's oeuvre. Edgeworth was a strident advocate for female domesticity, but her domestic ideal rested upon a rigorous education for women. As William Stafford notes, "Her aim is through education to make women rational, so that they can perform enlightened household management, take long views, calculate and regulate the expense of a family."[14] Narain further explains that, for Edgeworth, education was the means by which to teach women to embrace an exclusively domestic role and to enjoy doing so: "Properly educated women will see the value of domesticity with reasonable insight. While women do not have the privilege of choice, they do have the power to make the choice to follow men's choice and adjust to them."[15] However, participation in domestic networks did not preclude women from participation in scientific networks, especially at their points of intersection.

Women's education, for Edgeworth, was thus highly contingent upon its use value and its particular applications. That she endorsed a scientific education for women and girls is key. In *Letters for Literary Ladies*, for instance, scientific education is encouraged much more enthusiastically than one might think at first glance. In a significant passage, the more liberal father notes that the improvement in the style and presentation of scientific texts and the move away from "jargon" and "mystery," inspired by Francis Bacon and ostensibly practiced by his followers, was greatly beneficial to women seeking scientific knowledge. Edgeworth writes, "Our books of science were full of unintelligible jargon, and mystery veiled pompous ignorance from public contempt; but now, writers must offer their discoveries to the public in distinct terms, which every body may understand; technical language will no longer supply the place of knowledge, and the art of teaching has been carried to great perfection by the demand for learning: all this is in favour of women."[16] This idea, that contemporary developments in scientific writing have resulted in a more accessible type of science, is, of course, "in favour of women." However, the directive that scientific discoveries were to be elucidated in "distinct terms, which every body may understand," puts women's scientific comprehension essentially on par with that of male readers: it was not the case that women could not decipher "technical language" (which was clearly nothing but a cover for "pompous ignorance"), but rather that such language was obfuscating in general. Edgeworth further counters the notion that women were less capable than men of acquiring scientific knowledge by emphasizing how women learned and what resources they had available to them as learners. In essence, the "art of teaching" and the rejection of "fashionable publications" combined to make science accessible to an interested female audience.

Edgeworth continued her exploration of female education in the third portion of the text, "Letters of Julia and Caroline," a collection of letters that presents a typical narrative of one responsible young woman giving advice to a flightier companion. The ambitious aims of these letters are manifold, but I want to focus on their energetic debate about women and philosophy. While "philosophy" is used here to indicate intellectual inquiry in a broad sense, given the scientific specificity of the previous letters, it is easy to imagine scientific education as part of the appropriate "system of philosophy." Thus, these exchanges between the sentimental Julia and sober Caroline engage indirectly with the question of women's scientific pursuits, as Julia emphatically rejects Caroline's pleas for her to pursue a rational course of education by exclaiming, "What has a woman to do with philosophy?"[17] Julia's uncorrectable excesses (and her rejection of the idea of a "system" or programmatic approach to her education) ultimately lead her to a tragic death, in spite of Caroline's best efforts to rehabilitate her friend.[18] And, while Julia is clearly the irresponsible and ultimately "bad" character, the possibility of empathizing with such a spirited girl is markedly present in the text. Most compellingly, this concern with the propriety of a scientific education for women becomes coupled with its desirability. Edgeworth-as-Julia raises the tricky question of *why* a woman would want to pursue scientific education when it may bring her neither pleasure nor a husband; as she exclaims, "The graces flourish not under [philosophy's] empire; a woman's part in life is to please."[19]

Caroline attempts to defend intellectual inquiry as a way of finding true and enduring happiness in life, but Julia remains unconvinced. Julia eventually seems to garner a scant scientific knowledge, but only enough to make confused comments about Francis Bacon that pass for wit among her fashionable friends.[20] Perhaps Edgeworth thought of this dialogue as a means to counter such anxieties in her readers about whether to pursue philosophical studies of their own. To not do so, in this vignette, is to fail at cultivating morality and the ability to make the right decisions about momentous life occasions: Julia cannot decide which of two suitors would make a better husband, chooses the wrong man, and eventually leaves him for a life of dissipation in France that culminates in her death.[21] A thorough philosophical education for women is, in this narrative, literally a matter of life and death.

It is unclear how such enthusiastic advocacy for female education was received in the eighteenth century. The typically sympathetic *Monthly Review* critiqued *Letters for Literary Ladies* in 1796 for exceeding the bounds of an appropriate education for women, declaring, "We can never be persuaded that the state of the world would be improved by converting all our affectionate wives, kind mothers, and lovely daughters, into studious philosophers, or busy politicians."[22] Paul Keen notes further that the *Monthly Review*, while printing "with evident approval" comments on *Letters for Literary Ladies*, still hesitated to endorse texts that attempted to promote a woman to "eminent mistress of any single science or art."[23] Yet the cry to convert all women into "studious philosophers" was never what was intended in *Letters for Literary Ladies*: Edgeworth was abundantly clear about the limitations of what she expected women to attain. The male character in the epigraph to this chapter reminds his correspondent, "I do not desire to make [my daughter] a botanist, a mathematician, or a chemist."[24] There is no intent in this text to prevent or dissuade women from embracing their role as rational domestic creatures. In fact, Edgeworth qualifies the entire letter with the following disclaimer: "Do not, my dear Sir, call me 'a champion for the rights of woman'; I am more intent upon their happiness than ambitious to enter into a metaphysical discussion of their rights."[25]

Seemingly disavowing the ideologies of more assertive writers such as Wollstonecraft and Hays, Edgeworth—in the voice of her male character—dismisses the possibility of women as intellectuals, wishing only for their "happiness" within a domestic system that must be preserved. Yet the passage also suggests that Edgeworth believed scientific knowledge had a place within that domestic sphere, as a means of satisfying, enlightening, and pleasing women. While rational education might primarily be intended to prepare women to accept a domestic role, it is scientific education in particular that could create a happier, more fulfilled woman who circulated within that domestic system, as evidenced in Edgeworth's fiction.

Navigating Edgeworth's Scientific Curriculum

Edgeworth's interest in scientific pedagogy stemmed from a significant family tradition of scientific inquiry. As Marilyn Butler has outlined, Edgeworth's intellectual interests were indelibly shaped by her father's early and continuing influence, although Caroline Gonda has convincingly argued that the extent to which Edgeworth's father influenced her writing may have been a convenient fiction that enabled a more assertive intellectual program.[26] We know from Maria Edgeworth's letters and her publication of her father's memoirs in 1820 that her father conceived of himself as an inheritor of the pure passion for objective scientific inquiry that motivated Bacon.[27] He was also an active member of the Lunar Society.[28] An organization that dedicated itself in the late eighteenth century to pursuing passionately any number of scientific and philosophical projects, the Lunar Society boasted members whose work directly influenced Edgeworth herself, including Erasmus Darwin (whose *Zoonomia* and *Botanic Garden* she quoted in her educational works) and Thomas Day (whose *Sanford and Merton* she often recommended as a pedagogical tool for young boys). Edgeworth admired many of these thinkers and drew inspiration from their lives, experiments, and discoveries as she pursued a variety of writing projects throughout her life, from pedagogical volumes to popular novels. The Lunar philosophy of education also significantly influenced Edgeworth's approach to many of her projects, as she echoed in writing their desire that "education should be scientific in method and in content, fully understood by parents and teachers and given to all people of both sexes."[29]

While she was clearly an advocate for women's scientific education—both as children and as adults—Edgeworth's treatment of certain sciences in her early pedagogical works intersects in ambivalent ways with the prevailing opinions on the subject. For example, while botany and natural history were arguably the most popular pursuits for women at the turn of the century, Edgeworth most frequently and enthusiastically endorsed the study of chemistry, often at the expense of other sciences. In *Practical Education*, she devotes a specific section to chemistry for small children of both sexes, and in *Letters for Literary Ladies*, she emphasizes the study of chemistry as valuable to women because it made them more efficient in the kitchen and happier as housewives.[30] She writes, "Chemistry is a science particularly suited to women, suited to their talents and to their situation. Chemistry is not a science of parade, it affords occupation and infinite variety; it demands no bodily strength, it can be pursued in retirement, [and] it applies immediately to useful and domestic purposes."[31] This explanation is one of the most ardent endorsements for women's scientific education that Edgeworth makes in any of her early works, a statement whose passion is unmatched when speaking of any other sciences.

For example, Edgeworth mentions natural history only in passing in *Practical Education* as an appropriate subject of study for young people (as I will expand upon below). It is also included in one of her *Moral Tales* (1801), "The Good French Governess," with reference to the young character Herbert reading "Buffon, and several books, which contained good prints of animals."[32] In the same story, Herbert also learns valuable lessons through natural history, studying an ant for anatomical purposes and using a snail's speed to solve a math problem.[33] In "Forester," the extended friends and family of Dr. Campbell preview "an excellent cabinet of minerals," but the collection is not portrayed as anything other than a source of entertainment and curiosity.[34] However, Henry Campbell owns natural history books and conducts chemical experiments, in addition to keeping his own cabinet of fossils—evidence of his gentility and proper education.[35] Intriguingly, while the discipline of natural history appears periodically in her earlier works, it does seem confined to male characters: Edgeworth mentions elsewhere that she used "he" as the default pronoun for the sake of convenience, but her fully realized characters who engage in the study are often male.[36] The role of women in the study of natural history is thus largely unexplored in these early pedagogical works.

Botany fares a little better. In another one of the *Moral Tales*, "Angelina," the wise and kind preceptress of an "academy for young ladies" has her "young botanists" examine a "basketfull of bee orchis [orchids]," a curious plant (native to Britain) whose blossom mimics the appearance of a bumble bee.[37] As the only mention of an intellectual activity undertaken by the virtuous girls of Mrs. Porett's academy, the act of examining these singular plants is an important acknowledgement of the significance of botany to a holistic education for girls. It is also crucially combined with other female accomplishments in a comprehensive educational program: when Angelina comes upon the young ladies of the academy, they are learning to dance. And while botany has little narrative significance in "Angelina," it does serve to establish the contrast between the "right" type of pedagogy as practiced by Mrs. Porett and Angelina's flawed, sentimental education (as well as the equally flawed, aggressively philosophical education of her supposed friend "Araminta"). Finally, by making her encounter with Mrs. Porett's academy the beginning of Angelina's recuperation into an appropriate domestic mode of femininity, Edgeworth also reaffirms the ways in which scientific education can make women better, happier people.

However, Edgeworth also sometimes dismisses botany as a possible field of study for various reasons: in "The Good French Governess," Madame de Rosier does not purchase a microscope because "her pupils were not immediately going into the country, where flowers could be procured," and Charles Percival in *Belinda* is discouraged from sending too much information about a particular plant to Helena Delacour because the latter could not "find ditches and pools in Grosvenor square."[38] In this respect, Edgeworth is demonstrating an odd urban bias that could not have been inspired by her own residence in Edgeworthstown, which had a lawn with "many fine trees" adjacent to "rich grazing fields, a wood haunted by herons . . . and a great walled fruit-garden."[39] For some reason, she seemed to assume that the majority of her readers lived in a city or another location where the natural world was largely inaccessible (which was still untrue even in major urban centers like London, as William Curtis's *Flora Londinensis* demonstrates). Occasionally, Edgeworth even seems to disapprove of botany for other, more cultural reasons; in a passage from *Letters for Literary Ladies,* for example, she notes dismissively, "Botany has become fashionable; in time it may become useful, if it be not so already."[40] Given Edgeworth's habit of redacting inappropriate content for her young readers in *Practical Education,* perhaps she was at least partially influenced by the aspersions of sexual impropriety so often cast against botanizing women.

But this side-swipe—"if it not be so already"—contradicts the depiction of botanical study in *Belinda,* where botany represents important concerns and enables crucial narrative developments within the text even if often misapplied. Natural history also plays a significant role in *Belinda,* as the nearly ideal Percival family comes to embody Edgeworth's ideas from *Practical Education* about the significance of the study of nature to a cohesive and healthy family environment. Likewise, the successful suitor for Belinda's hand, Clarence Hervey, shows evidence of a solid scientific education. Where Edgeworth only sketches out a vision for scientific study in *Letters for Literary Ladies,* she fully embodies that vision in *Be-*

linda, in which scientific knowledge and references permeate the entire text. Furthermore, it is Belinda herself who mediates nearly all of the scientific knowledge in the novel, making her both an ideal natural philosopher and, as many critics have pointed out, a mouthpiece for Edgeworth's own ideas about education. And while Deborah Weiss has usefully interpreted Belinda as a "true . . . female philosopher" who represents those positive qualities associated with Enlightenment thought (in contradistinction to Harriet Freke), I am expanding upon this premise to emphasize Belinda's role as a representative of beneficial naturalist study.[41]

Edgeworth's ideas about scientific education were derived from broader theories about child-rearing and education as espoused by her family. Several critics have noted the significance of experimentation to Edgeworth's pedagogy, including Narain, Ruth Watts, Catherine Toal, and Julia Douthwaite, who have all highlighted the idea of "experiment" as central to her educational praxis.[42] Susan Manly explains this philosophy in the context of her educational works for children: "Edgeworth's educational programme . . . discards theoretical creeds in favour of experimental approaches, and . . . seeks to lead children out into the wide spaces of exploration and invention, using their immediate environment and everyday activities."[43] The best way to teach young children of both genders about science, Edgeworth believed, was through hands-on experimentation. Most important for women as educators of children and as individuals seeking their own scientific edification, experiments for Edgeworth would be ideally conducted in the home. Her interest in experimentation was further differentiated from that of professional male scientists in multiple ways, most significant of which was her disapproval of the element of spectacle that many public demonstrations like those put on by the Royal Society encouraged. For the men undertaking such experiments, as many historians of science and critics have noted, scientific experimentation was already performative, whether conducted in private chambers or public venues.[44]

Thus, for Edgeworth as a writer and a pedagogical theorist, it was also vital that experimentation be stripped of any mysterious excitement or spectacular effects. Experimentation was to be done clinically, practically (a concept she evokes frequently, and not just in her titles), and with many explanations along the way to demystify what was occurring as much as possible. Nicole M. Wright's study of *Belinda*'s depiction of chemistry posits that Belinda's significance in the novel is generated by her role as debunker of scientific mystery. Wright explains that the novel contains a number of scenes in which "the application of scientific methodology resolves the obfuscation that is perpetuated through superstition and trumped-up spectacle," such as when Belinda exposes Juba's terrifying "obeah woman" as nothing more than glowing phosphorus smeared on the wall.[45] While these scenes highlight Belinda's role as a figure that represents proper education, they also "indicate Edgeworth's determination to establish experiment as a central, non-sensationalized aspect of a scientific education."[46]

Belinda's function as a scientific influencer can also be used to examine broader themes of scientific study that are present throughout the novel—namely, the Percival family's positive identification with natural history and Lady Delacour's problematic association with botany. I argue that Belinda serves as the moderator of scientific knowledge throughout the text: approving of and enabling the Percivals' (and Helena Delacour's) continued study of natural history; regulating and correcting Lady Delacour's misunderstanding of botany; and even demystifying Harriet Freke's mischievous deployments of chemistry (as elucidated in Wright's article). In this way, Edgeworth exposes the problems inherent in the partial knowledge of science she worried about in *Practical Education*, while also noting approvingly when women get scientific education right. Further, by incorporating different fields of scientific study for women, Edgeworth is also evaluating the usefulness of such areas of inquiry to the contented domestic existence for which women were likely destined.

The scientific underpinnings of the novel, when read in conjunction with her other pedagogical publications, expand our understanding of Edgeworth as an author deeply invested in exploring and promoting female networks of knowledge. Few other critics besides Wright have touched upon Edgeworth's scientific interests in *Belinda*. Kathryn Scantlebury and Collette Murphy have written a brief summary of the scientific themes in *Belinda*, primarily with an eye to highlighting her interest in creative pedagogy.[47] Richard Holmes has also addressed the scientific components in her oeuvre, although he frames her interest as merely manifesting a fictionalized version of male scientists and their work.[48] In contrast, I present a sustained study of Edgeworth's scientific curriculum that argues for her aims as both more ambitious and more consistent than these previous critics have acknowledged.

Studying Nature, Learning Virtue

Edgeworth believed that the study of natural history was an excellent way to introduce children to science as a field of knowledge because it required them to engage thoughtfully with objects that already interested them. She writes in *Practical Education,* "Natural history is a study particularly suited to children; it cultivates their talents for observation, applies to objects within their reach, and to objects which are every day interesting to them. The histories of the bee, the ant, the caterpillar, the butterfly, the silk worm, are the first things that please the taste of children, and these are the histories of realities."[49] Of course, the "histories" of these many small creatures was also of interest and value to engaged adult women, as evidenced by the studies of previous chapters. This passage may very well be directed in part at mothers who sought to educate themselves along with their young children, the implication being that natural history is simply a more accessible science for beginners than some others. (And, like the Lunar Society members, Edgeworth believed in educating individuals of both sexes in the sciences.) Note also the final clause of the passage, in which Edgeworth defends natural history as "the histories of realities." Once again, she places science in implied opposition to ostensibly fanciful genres like romance, which fellow novelist and critic Clara Reeve so casually dismissed as "what has never happened, nor is likely to happen."[50] This insistence upon reality and practicality permeates all of *Practical Education* and *Belinda,* where fancies and imagination must be harnessed and redirected in pragmatic ways—both in children and adults. And though perhaps older readers had no need of cultivating their observational skills or learning to pay attention, it is also a very real possibility that such practices could benefit adolescent women as well, especially those at an impressionable age whose vulnerability to social missteps (or worse) was significant.

In addition to prescribing natural history as an appropriate field of study for young women and girls, Edgeworth provides details about the type of equipment and tools required to pursue such studies. A garden and a curiosity cabinet, for example, are specific items among the useful components in the study of natural history.[51] The former was designed to encourage both industry and horticultural skills and the latter intended to permit the collection of objects of fascination and edification. Her specific interest in natural history is usefully elucidated in "The Good French Governess," from *Moral Tales,* in which the governess Madame de Rosier takes over the mismanaged education of four young children. Madame counters the fashionable education given to the young girls and the neglected education of the little boy by introducing them to rational pursuits and intellectual opportunities. She begins by curtailing their competitive natures and impetuous behaviors and then subtly proposing the study of scientific subjects (at an age-appropriate level like the kind elucidated in *Practical Education*). She does so with the visit to a particular shop, a "warehouse for toys, which had been lately opened, under the direction of an ingenious gentleman, who had employed proper workmen to execute rational toys for the rising generation."[52]

In itemizing potential purchases in this "rational toy-shop," Edgeworth elaborates on the types of scientific equipment appropriate for young learners, including cabinets for "young mineralogists," "a botanical apparatus," and a "magnifying glass."[53] While the "botanical apparatus" is likely purposefully left vague (most tools required to study botany could be found around the home and, in conjunction with that magnifying glass and specimen cabinet, would be easily accessible), there is a kind of specificity to this list that bespeaks Edgeworth's significant experience in teaching science to young children. She further asserts that children should also have access to "cheap microscopes" that are both entertaining and informative, revealing "a world of new delights" to curious viewers.[54] By the end of the eighteenth century, inexpensive and portable microscopes were readily available for purchase, and while they were not likely to be owned by working-class families, they were no longer a luxury item, either.[55] For families that embraced the new domestic model of education like the Edgeworths, the value of introducing young children to the basics of scientific inquiry before any period of formal schooling (for boys) or alongside other essential competencies like sewing (for girls) was tremendous.

Crucially, in the model of education proposed by Edgeworth in "The Good French Governess," both boys and girls are included in this "world of new delights." So, too, presumably, is their mother, as the person primarily responsible for educating her young children. While most mothers of young children at the turn of the century could have had access to the wealth of scientific material available from the popular press, her own knowledge would have depended significantly upon the attitudes of her parents or guardians. Likewise, her decision to educate her own children in the sciences arguably depended even more upon her husband's approval (or lack of interest). Yet, as the emerging conceptualization of domestic motherhood attests, it was the woman's responsibility to provide the best education possible for her children: perhaps even in spite of an indifferent or antagonistic husband. For Edgeworth, this required a certain level of scientific familiarity that, if not already possessed from childhood, needed to be acquired as an adult. In *Letters for Literary Ladies*, she writes, "Ladies have become ambitious to superintend the education of their children, and hence they have been induced to instruct themselves, that they may be able to direct and inform their pupils. The mother . . . must have a considerable portion of knowledge."[56] With regard to natural history, this was assumed to be a simple task for most sensible women, who only had to direct their pupils to observe carefully and thoughtfully and to allow the children's natural curiosity to thrive. Of course, as becomes evident in *Belinda,* not all women are sensible, and not all mothers have the patience and discernment to properly educate their children in the sciences.

It seems appropriate that Edgeworth published *Moral Tales* and *Belinda* in the same year. By 1801, her ideas about scientific practice from *Letters for Literary Ladies* and *Practical Education* had been further developed, influenced by a few more years of helping to raise her younger siblings and by her own evolving thinking about women's education. Consequently, in "The Good French Governess" and *Belinda,* she offers what she then believed to be the best way to encourage scientific education among children and women. Both texts rely upon the guidance of an ethical and responsible mother figure (the governess Madame de Rosier and Lady Anne Percival, respectively), the deployment of experimentation as the primary means of discovering knowledge, and the insistence upon practicality in experimentation (rather than the pursuit of theoretical knowledge). Where "The Good French Governess" gestures toward appropriate scientific education in miniature, *Belinda* enacts it on a broader scale with greater social consequences.

Natural History and the Novel

The Percivals are essential characters in *Belinda*'s elucidation of appropriate scientific inquiry for women. While they make the significant mistake of endorsing Mr. Vincent as Belinda's suitor (unaware that he has a chronic gambling problem and, as Kathryn Kirkpatrick points out, ignoring his problematic colonial interests), their pedagogical practices position them as a model family for the Delacours and readers of *Belinda* to follow.[57] Douthwaite notes that eighteenth-century readers would have recognized the Percivals as an ideal family from the outset, explaining, "The name *Percival* itself signified progressive science to many readers. Thomas Percival was a well-known chemist, philanthropist, and peer of luminaries such as Joseph Priestley, Benjamin Franklin, and René Antoine de Réaumur."[58] In this intellectual legacy, we see the Percivals not only following these illustrious philosophers, but also the

Edgeworths' own pedagogical program. Heeding the advice from *Practical Education* that "an early knowledge of the first principles of science may be given in conversation, and may be insensibly acquired from the usual incidents of life," all of the Percival children are active participants in the intellectual life of the family.[59]

While staying at their home in Oakly-park, Belinda finds herself "in the midst of a large and cheerful family," observing that "there was an affectionate confidence, an unconstrained gayety in this house."[60] More important, even the youngest members of the family "seemed to form part of the society, to have some share and interest in the general occupations or amusements [of the entire family]."[61] Their questions are encouraged, their speculations welcomed, and their learning a process of trial and error. Belinda "heard none of that continual lecturing, which goes forward in some houses, to the great fatigue and misery of all the parties concerned," but instead found that both Mr. and Lady Percival provided their "knowledge of the world" freely and comfortably.[62] The children themselves were not without individual personalities and skills, either: Edgeworth notes that "one of the boys was fond of chemistry, another of gardening; one of the daughters had a talent for painting, another for music."[63]

The harmony of scientific and literary inquiry found in the Percival home closely echoes the atmosphere of the Edgeworth home, a social space that sympathetic visitors to Edgeworthstown recognized as pedagogically ideal. In 1810, Charles Kendal Bushe wrote approvingly of the system of education that had been established within the Edgeworth household, the description of which echoes Edgeworth's fictional depiction of the Percivals from nearly a decade prior. He writes:

> Society in that house is certainly on the best plan I have ever seen . . . There are many young Edgeworths male and female all of promise and talent and all living round the same table . . . In such a Society you may suppose Conversation must be good, but I was not prepared to find it so easy. It is the only set of the kind I ever met with in which you . . . [fall] imperceptibly, into literary topics, and I attribute it to this that in that house literature is not a treat for Company upon Invitation days, but is actually the daily bread of the family.[64]

Although such a visit occurred nine years after the publication of *Belinda*, it is easy to believe that such a lively system of education and conversation had continued to develop over time. By replicating in fiction the pedagogical environment that influenced her own life, Edgeworth affirms the value of intellectual inquiry to a happy domestic existence.

This idea that knowledge, both literary and scientific, is the "daily bread" of the Edgeworths and the Percival family is made evident in the young Percival children's experiment with a goldfish bowl. Upon Clarence Hervey's first visit to Oakly-park, we find the Percival children crowded around a fishbowl, engaging with their household pet in an investigatory manner: "One of the little boys flipped [tapped] the glass globe, and observed, that the fish immediately came to the surface of the water, and seemed to hear the noise very quickly; but his brother doubted, whether the fish heard the noise, and remarked, that they might be disturbed by seeing or feeling the motion of the water, when the glass was struck."[65] While this is surely a playful engagement with the natural world, it is precisely the harnessing of such playfulness into teaching moments that embodies the Edgeworths's pedagogical project. Empiricism is key here: as Wright notes, "The children find occasions in their everyday experience," like the inspection of a fishbowl, to improve "their capacity for informed judgement" and rational thought.[66] Moreover, it is in response to such a teaching moment that *Belinda* concerns itself directly with natural history. Dr. X—, a wise friend of the Percivals who plays multiple corrective roles in the novel, raises the point that the question of fish being able to hear had been a subject of some debate in contemporary natural history. He "observed, that this was a very learned dispute, and that the question had been discussed by no less a person than the abbe Nollet, and he related some of the ingenious experiments tried by that gentleman to decide, whether fishes can or cannot hear."[67]

Jean-Antoine Nollet, a professor of experimental physics at the College of Navarre, was primarily known for his experiments on electricity and even famously engaged in a transatlantic debate with Benjamin Franklin on the subject.[68] He was also responsible for first discovering the phenomenon of osmotic pressure, an important development in understanding human and plant biology.[69] He was, however, also known for work on the theory of sound, conducting several experiments on the transmission of sound through various substances and disseminating his results in a massive, multivolume publication (ca. 1743). This publication, *Leçons de physique expérimentale,* presented not only his findings but also the elucidation of his many experiments on the transmission of sound in Leçon XI, "Du Milieu qui transmet les sons" (found in vol. 3). Nollet first describes removing the air from an experimental vessel by submerging a leaden device and then rapping on the vessel to see if the sound carried through a vacuum; he finds in the negative ("un timbre qui fait ses vibrations dans le vuide, ne les peut communiquer à rien: par conséquent, puisqu'elles n'operent le son que quand elles se transmettent, elles doivent se passer dans le vuide avec un profond silence").[70] Then, a second experiment involving water found that sound did transmit, but much more slowly and at a lower volume than through air ("il es vrai que tous ces sons étoient sort affoiblis").[71] Ultimately, Nollet concludes that, with no proof of fish having ears, or no idea of how they perceive exterior stimuli, the fact that sound transmits through water does not illuminate the ability of fish to hear.[72]

While Edgeworth does not include all of this information in *Belinda,* the fact that she was familiar enough with the work to attribute such knowledge to the character of Dr. X— speaks volumes about her own scientific education. This particular publication, although first printed in 1743, was reprinted several times and may have been among the scientific texts owned by Richard Lovell Edgeworth, thereby providing Edgeworth herself access to it. Extant letters from Edgeworth also indicate that her father collected issues of *Philosophical Transactions* for decades, further suggesting that while she might not have read the relevant issues herself, their contents were likely shared with the family as part of the Edgeworths' conversational style of education.[73]

The inclusion of Nollet as a source of expertise on natural history, however, accomplishes more than simply attesting to Edgeworth's own knowledge of the subject. It ratifies the Percival children's interest in the subject as more than the idle curiosity of children and proves to the reader and to Hervey that they are astute enough to ask questions that even the great scientists of the Royal Society and the French Royal Academy of Sciences thought to pose. It also underscores the Percival household as an ideal space for scientific inquiry, where natural history is pursued both harmoniously and appropriately. Further, Nollet's theories of sound in water serve to illuminate additional perspectives on natural history: Belinda's measured and appropriate position, and Lady Delacour's dismissive and ignorant perspective. For example, when Hervey attempts to proudly pass along his newfound information about the Abbé Nollet (note here that the adults in the room learned just as much as the children during that Oakly-park sojourn) to Lady Delacour, she exclaims dismissively, "Deep in philosophy, I protest! . . . What is this about the transmission of sound in water? Ha! whence come these pretty gold fishes?"[74]

Lady Delacour represents a person ignorant of natural history, an individual whose education has not only been confined to the fashionable world (and those devious fashionable publications, as Heather MacFayden has pointed out) but also completely neglected scientific subjects.[75] "Philosophy" is something to be objected to as unfashionable, unimpressive, in popular society; Lady Delacour is much more immediately concerned with what impresses her audience, just like her erstwhile companion Harriet Freke or Julia in *Letters for Literary Ladies*. Particularly given the visual and cultural associations with goldfish—they serve as signifiers of wealth and the desire for exotic commodities—I also read Delacour's spark of interest in the fish, however fleeting, as condemning her materialist impulses.[76] However, philosophy in this case also proves to be what reunites Helena Delacour with her mother. Initially an anonymous gift-giver who passes along the goldfish, Helena later admits during a much longed-for visit to her estranged mother that she had actually won them from Mrs. Percival because she, out of all the children in the Percival household, was able to reproduce the best explanation of the Abbé Nollet's theories.[77] Although the initial passage mentioned only male children, and Belinda's observations of the Percival family note only a son interested in natural history, it is the young woman here who best comprehends advanced natural history and experimental philosophy. Further, it is Belinda's generous and affectionate understanding of Lady Delacour's true character that inspires her to encourage Helena in her plans to give the fish to her mother (in the hopes of replacing Marriott's noisy macaw).[78]

Nollet's theories thus contribute to the education of a wide cast of characters: the eager young Percivals, the shy and developing Helena, and the recovering Lady Delacour. And, by assuming responsibility of the passage of the goldfish, Belinda becomes the conduit through which the appropriate knowledge about natural history is transmitted: Hervey's attempted injection of the details of Nollet's theories is rejected, but Helena's gentle explanation of them is accepted and valued by wiser characters. In this triangulation between Helena, Belinda, and Lady Delacour, we observe most clearly Edgeworth's investment in women's scientific education in particular. Readers do not especially care whether Hervey knows about goldfish, and it is not important that Dr. X— relays this information. What is important is that natural history ultimately serves the function of beginning to reunite an estranged mother and daughter.

This linkage between Helena Delacour and natural history extends further in the novel. Prior to Lady Delacour's rehabilitation, she receives a letter (one of many) from her daughter that she feigns not to have read. Belinda inquires about the letter, at which point Lady Delacour admits that she has read enough of the contents to forgo finishing the missive: "No, I saw something about the Leverian Museum, and a swallow's nest in a pair of garden shears; and I was afraid I was to have a catalogue of curiosities, for which I have little taste, and less time."[79] While providing additional proof of the Percivals' sensible education of their children, this visit to the Leverian Museum also stands out in the novel as a moment where natural history serves as a generative force, when directed appropriately by Belinda. In fact, this scene is one of the earliest in which Lady Delacour expresses her repressed affection for her daughter, and in which Belinda begins her efforts to reconcile them—prompted exclusively by this letter.

The reference to the Leverian Museum is only one of the many junctures in the novel at which Edgeworth appeals to her readers' own knowledge to expand the imaginative scope of the narrative. As eighteenth-century readers would have known, Ashton Lever was a consummate collector who had acquired through great expense a vast assemblage of natural artifacts, many of which were from James Cook's three voyages to exotic parts of the known and unknown world.[80] When they became too numerous to contain at home (the original collection held an estimated twenty-six thousand objects), he founded the Leverian Museum in 1774 and opened it to the public for the fee of half a guinea.[81] Artifacts included fossils, shells, fungi, botanical specimens, and preserved animal specimens including insects, reptiles, and fish.[82] There were also live specimens, including monkeys and "monsters," which were kept in separate rooms to avoid offending or startling female visitors.[83] Lever's original collection was transferred in 1786 to James Parkinson, who inherited the entirety of the collection from his deceased wife's winning lottery ticket (in financial distress, Lever had "obtained an Act of Parliament, allowing him to dispose of the whole by lottery").[84] Upon Lever's death, Parkinson moved the collection from its original location in Leicester Square, and it remained on display on Blackfriars Road until 1806, expanding frequently and with much fanfare by adding new, exotic specimens, including an elk, a kangaroo, and a condor.[85] By the time the Percivals and Helena were visiting, the Leverian Museum was at the height of its popularity, with curious visitors and tourists of all kinds, many of whom were interested in natural history and found the opportunity a priceless experience.

We do not know if Edgeworth herself ever visited the Leverian Museum. She spent little time in London and does not mention a visit in her letters; it is perhaps easiest to conclude that she did not (although she did visit the natural history museum associated with Trinity College during one of her many stays in Dublin, much later, in 1815—the topic was clearly of interest to her).[86] She may have, however, read *A Companion to the Museum*, a guide to the Leverian Museum published in 1790 (and compiled by George Shaw, an assistant keeper of natural history at the British Museum) which gave a room-by-room overview of the contents.[87]

The museum was also prominent in popular culture, and extant publications indicate that other women engaged with this wealth of natural history in a variety of ways. Lady Ellenor Fenn, for example, wrote and published two texts that were designed as guides for those walking through the museum, *A Short History of Insects* and *A Short History of Quadrupeds*. Although both works were undated (making their precise time of publication difficult to pin down), they were likely to have been available to a broad audience, due to the museum's immense popularity.[88] Like several of the natural history books that the Percivals owned, and that many other women in the period possessed and consulted regularly, these volumes were presented with detailed hand-colored drawings accompanying each topic, with both English and Latin names of each species.[89] The Leverian Museum was also of interest to artists, particularly as a study of exotic objects that most individuals would not have otherwise been able to access. Sarah Stone, a natural history artist and engraver, was commissioned by Lever to produce sketches and paintings of objects in the museum as part of a promotional campaign in the 1770s and 1780s.[90] She produced striking, hand-colored images of cultural artifacts, natural objects, and many items from Polynesia, the Indies, and North America. It seems likely that Edgeworth herself knew how important the Leverian Museum was to intellectual and artistic women like Fenn and Stone, making Helena's visit there all the more significant. By drawing upon her readers' familiarity with popular scientific culture, Edgeworth is able to give Helena a lively off-page life of scientific inquiry that underpins the networks of female exchange within the novel itself.

In addition to such museum visits, natural history books were a large part of the Percivals' scientific education, as was continued engagement with live animals. During Belinda's trip to the bird fancier's to replace Marriott's macaw, she accidentally encounters Lady Anne Percival, her children, and Helena Delacour visiting the same shop (an accident leading to their fortuitous introduction and an invitation to visit Oakly-park). The children become enchanted by a little goldfinch that has been trained to perform tricks, and they exclaim, "O! mamma . . . this is the very thing that is mentioned in Bewick's History of birds. Pray look at this goldfinch, Helena—now it is drawing up its little bucket."[91] In this passage, Edgeworth is demonstrably drawing upon her very current knowledge of natural history and popular publications in the field. Thomas Bewick was an engraver and natural history author who published multiple texts on the subject.[92] The Percivals are most likely referring to *History of British Birds*, published in Newcastle in 1797 by Ralph Beilby, Bewick's mentor and employer in the engraving business.[93] The engraving of a goldfinch is found partway through the first volume with the cited explanation: "In a state of confinement [goldfinches] are much attached to their keepers, and will learn a variety of little tricks, such as to draw up small buckets containing their water and food, to fire a cracker, and such like."[94] It seems evident that Edgeworth drew this scene directly from the volume in question: she mentions Bewick's work specifically in a letter to Sophy Ruxton in 1805 about her goddaughter, Sophy Edgeworth, indicating that she utilized his books in educating the younger children in natural history.[95]

In *Belinda*, the knowledge of animals and natural history as demonstrated by the Percival children and Helena Delacour also serves as a foil for the adult lack of knowledge about botany. Readers know to approve of the children's active spirit of inquiry as they approach both goldfinch and goldfish, but Lady Delacour's attitude toward botanical knowledge has usually been elided. Crucially, there is a jarring contrast between Helena's pursuit of naturalist knowledge and Lady Delacour's seeming indifference to all sorts of scientific knowledge that reveals Edgeworth's narrative strategy for promoting scientific study among women of all ages.

Edgeworth effectively establishes this contrast through the narrative and physical movement of the goldfish. After Helena has sent the goldfish along to her mother as a silent pet who would not disturb her during her (psychosomatic) illness, Charles Percival interrupts Belinda's letter-writing to ask, "Will you send this in your letter to Helena Delacour, and tell her that is the thing that gold fishes are so fond of? And tell her that it is called lemna, and that it may be found in any ditch or pool."[96] Lemna, or duckweed, is a broad genus of water plants that could be found across Europe and around the world, but the specificity of which type of lemna is not important here.[97] What is of significance is that both Charles Percival and Belinda know and recognize the Latin name for a very common plant. Lady Delacour, upon receiving the goldfish and the plant, has a very different reaction:

> Marriott brought in her gold fishes; some green leaves were floating on the top of the water in the glass globe.
>
> "See, my lady," said she, "what miss Portman has been so good as to bring from Oakly-park for my poor gold fishes, who, I am sure, ought to be much obliged to her, as well as myself." Marriott set the globe beside her lady, and retired.
>
> "From Oakly-park! And by what name impossible to pronounce must I call these green leaves, to please botanic ears?" said lady Delacour.
>
> "This," replied Belinda, "is what 'th' unlearned, duckweed; learned, lemna, call', and it is to be found in any ditch or standing pool."[98]

The readers already know from Charles Percival that the goldfish food is called lemna, but Belinda is given the task of relaying this information to Lady Delacour and thereby reinforcing her superior scientific knowledge. It is not only Belinda's awareness of the Latin and vulgar names of a common plant that matters here: it is also vitally important that Lady Delacour's reaction to the plants (and the fish) is one of light-hearted dismissal. By critiquing that "name impossible to pronounce"—echoing Jacson's young character Harriet in her worry over correct pronunciation of botanical names—Delacour flippantly categorizes all of botanical science as useless to her. The idea of pleasing "botanic ears" is also a bit of snark, intended to downplay her ignorance while teasing anyone with serious botanical pretensions. And, though Belinda's remark is an affectionate one, indicating her recognition and approbation of her companion's gradual shift toward becoming a better person and mother, Belinda's response also contains a sorely needed corrective to Lady Delacour's bad ideas about botany.

Edgeworth's Plants

Botany evidently played a significant role in Edgeworth's personal life. Two sisters of her father's fourth wife, Frances Beaufort, anonymously published texts on botany and natural history in 1819, entitled *Dialogues on Botany and the Physiology of Plants* and *Entomology*.[99] Edgeworth approved of these intellectual ventures and even went so far as to send copies of the works to Rachel Mordecai Lazarus in America as part of their thriving exchange of botanical ideas and specimens—specimens including a hummingbird's nest, grasses and mosses, choice bird feathers, a variety of seeds, a "soda apple," alligators' eggs, unspecified berries, dried crocus bulbs, and even a Venus flytrap (which, sadly, never made it to Ireland alive).[100] She also reports many years of collecting and preserving plant specimens, including everything from the "quaking grass" (*Briza*) she loved to gather in decorative bouquets to an impressive variety of exotic American plants and seeds sent to her from Lazarus.[101] Over the course of her life in Ireland Edgeworth also cultivated a number of gardens with increasingly complex designs and expensive plants (several seeds and bulbs for which were regularly provided by Lazarus until her death in 1838), indicating an avid interest in horticulture if not in botany per se.[102] She also periodically corresponded with gardeners at the botanic gardens in Dublin and London, often sending them seeds and roots for identification or in the hopes that their efforts at growing American plants would prove more successful than her own attempts at Edgeworthstown.[103] Finally, she took an active interest in the scientific career of one of her youngest brothers, Michael Pakenham, whose position in the Indian Civil Service enabled him to pursue his own botanical interests, be elected a member of the Linnean Society in 1842, and publish a number of works on botany from 1851 onward.[104]

Further, Edgeworth's frequent interest in and repeated citations of Erasmus Darwin's *Botanic Garden* indicates that she was deeply invested in the pedagogical possibilities of creative engagements with vegetality. She found Darwin's volume so edifying that she directed even young children to read the text—a contrast with many conservative commentators in the period. In *Practical Education,* she mentions a ten-year-old child reading "parts of the second volume of Dr. Darwin's Botanic Garden," referring specifically to *The Loves of the Plants,* by far the more botany-based volume of the two.[105] The evidence that the child reads only "parts" suggests that the Edgeworths employed their favorite tactic of redacting suggestive passages from a text, but the fact remains that this highly contested botanical volume was a staple in the Edgeworths' pedagogical arsenal. This may have been in part due to the family's intimacy with Darwin himself (through Richard Lovell Edgeworth's involvement in the Lunar Society), but that does not entirely explain the many references to the text throughout Edgeworth's corpus. For Edgeworth, *The Botanic Garden* provided valuable and accurate information about the natural world, including extensive explanatory footnotes that contextualized the fanciful imagery. Such imagery also extended beyond the depictions of sensualized plants and included elements of natural history: as she explains in a chapter on "Grammar and Classical Literature" in *Practical Education,* "In Dr. Darwin's 'Botanic Garden' there are some beautiful poetic allusions to ancient gems and ancient fables, which must fix themselves in the memory or the imagination of the pupil."[106] Here, it is not only that the poem is valuable for its exposition of geology—both real and imagined—but also that such information can teach readers valuable lessons about important moments in history.

Edgeworth wholeheartedly encouraged parents to employ botany to teach their children about organizing schemes and patterns. In *Practical Education* she notes, "Botany and natural history, though they are not pursued as sciences, are, notwithstanding, the daily occupation and amusement of children, and they supply constant examples of classification."[107] As a result, much of the emphasis in both *Practical Education* and *Belinda* is on ways of processing stimuli and categorizing information in order to make appropriate decisions—whether as a child learning that a fire melts ice (a sample exercise in *Practical Education* to teach children introductory principles of matter), or as an adult discovering that a ghostly image on a wall is nothing more than a chemical specter (as in the example of Juba's "obeah woman").[108] Therefore, while Edgeworth does not engage directly with Linnaean taxonomy to the extent that the other women of this study did, she is evidently deeply invested in the underlying principles of order and control over the natural world.

In the above passage from *Practical Education,* botany is valuable not as science per se but as a discipline that trains children in learning how to think. In *Belinda,* however, Edgeworth utilizes the creative flexibility of the novel as a genre to explore the ways in which the lack of botanical knowledge can be a significant disadvantage in a woman's life. Specifically, the elasticity inherent in the form of the novel (a genre in which eighteenth-century authors envisioned the impractical adventures of individual atoms, African ostrich feathers, and enterprising servant-women, to name only a few) allows her to animate scenarios of disruption with the express purpose of directing Belinda to soothe and heal those disruptions.[109] By giving a woman the power to reconcile difficult scenarios, not through her goodness of temper and inspirational morality, but through her intellect and wit, Edgeworth also utilizes the novel form to promote women's education. While the academic bent of *Practical Education* and the truncated scope of the *Moral Tales* limit the imaginative possibilities of such forms, Edgeworth takes full advantage of the novel's latent potential to give life to unlikely circumstances in order to advance her pedagogical agenda.

Botanical Sins and New Beginnings

Edgeworth's pedagogical works frame botanical study as a generative field of inquiry for women; *Belinda,* on the other hand, envisions outrageous scenarios in which botanical knowledge goes awry. The relationship between botany and adult women is wound through the history of Lady Delacour, stretching back to many years before she met Belinda and concluding with her rehabilitation near the end of the novel.[110] Where we might expect an accomplished and affluent woman to have been afforded an excellent education in all appropriate subjects—including botany—Delacour possesses only a partial knowledge of this popular topic. Amy M. King refers to Delacour's partial knowledge of plants as "botanical fashion," but such a categorization does not encompass the full scope of her involvement in botanical activities and ultimately elides Edgeworth's palpable investment in scientific strategies of women.[111] More egregious than Delacour's ignorance is her vicious deployment of botany: she uses such imperfect knowledge only to enact petty revenge on her enemy Mrs. Luttridge. As MacFayden has outlined, Lady Delacour's intellectual presentation is based upon "[misusing] her literary knowledge and skill to support a rapidly altering sequence of personas whose novelty and daring enable her to maintain her public preeminence."[112] In other words, her education has been consistently devoted to nothing more than idleness and coquetry, as evidenced by her youthful refusal to settle down with Mr. Percival (her first and preferred suitor) and her continued adult dissipation. Her desire for revenge, then, is on par with her superficial and petty character. However, as readers of the novel well know, such revenge literally backfires on Lady Delacour, prompting her extended illness and genuine misery. Luttridge's animosity and equally competitive nature result in both women having "rival routs, rival concerts, rival galas, [and] rival theatres," but it is one conflict in particular over a specific wintertime gala that begins Delacour's tragicomic arc.[113]

As Lady Delacour is feverishly retelling her "history" to Belinda early on in the novel, she mentions a singular event in her relationship with Mrs. Luttridge that substantially fed their ongoing acrimony—the aloe affair. Delacour explains to Belinda that both ladies were putting on galas on the same evening, and Delacour sought desperately for some way to outdo her rival. In the midst of such machinations, she lights upon a botanical idea:

> Mrs. Stanhope's maid had a lover, a gardener, who lived at Chelsea; and the gardener had an aloe, which was expected soon to blow. Now, a plant that blows but once in a hundred years is worth having. The gardener intended to make a public exhibition of it; by which he expected to gain about a hundred guineas. Your aunt Stanhope's maid got it from him, for me, for fifty; and I had it whispered about, that an aloe in full blow, would stand in the middle of one of lady Delacour's supper tables.[114]

Lady Delacour's idea to utilize an impressive plant to attract crowds to *her* party instead of Mrs. Luttridge's has multiple social and botanical valences. While a purportedly rare plant that "blows [blooms] but once in a hundred years" would be a sight worth seeing—so much so that the gardener expected a significant sum of a hundred guineas for exhibiting the plant—an aloe is neither especially rare nor does it bloom so infrequently (although this might vary significantly depending upon the exact variety of aloe).

But who is ignorant here of the botanical cycles of the aloe plant? We might think first of Edgeworth: perhaps her botanical education at this early point in her life had been insufficient, given that the bulk of the evidence of her investment in botany points to an interest in collecting and cultivating, rather than categorizing or dissecting. But when the majority of her other scientific information is correct (including her knowledge of at least some botanical Latin), why would she be ignorant on this point? Would Edgeworth fall victim to her own caution against assuming expertise from merely having been "taught something of botany"?[115] The next candidate might be the gardener, who would be expected to know his plants more intimately and, in this case, would be woefully remiss in his duties. Given Edgeworth's class-bound prejudices—in the chapter on servants in *Practical Education,* she dismisses an entire class of people as stupid and immoral—it is plausible that she views a gardener (who was, after all, inferior to a professional botanist) as a figure whose craft was ultimately flawed by ignorance. However, if the gardener is ignorant, then, more important, so are Lady Delacour and all of her fashionable friends. Delacour is willing to splash out a significant sum of money to snatch the impressive aloe away from the gardener and put it on her own table, all while being unaware that the plant was not particularly impressive. As fashionable people (and fashionable readers), Delacour and her guests represent flawed intellectual perspectives who seemingly have only that partial knowledge of botany.

Edgeworth's choice of the aloe as a figure of fascination is an intriguing one. Setting aside that the sixth edition of Philip Miller's *Gardeners Dictionary* (1752) lists thirty-eight different species of aloe, I suspect that she refers to *Aloe vulgaris* (today, *Aloe vera*), which does not grow naturally in Britain but was a well-known acclimatized species and produces yellow or orange blossoms.[116] A visually striking species, *Aloe vulgaris* was a plant that botanists had found captivating for generations. Prior to *Belinda*'s publication, Edgeworth may have read John Abercrombie's popular *Every Man His Own Gardener*—in its sixteenth edition by 1800—which had an expansive catalog of hothouse plants, including many varieties of aloe.[117] It also seems that Edgeworth's idea for an aloe that "blows but once in a hundred years" came from an urban legend. In the lengthy description of aloe growing and blossoming habits found immediately after the catalog of different aloe species in *The Gardeners Dictionary*, Miller writes, "Here I cannot forebear taking notice of a vulgar Error or two relating to the large *American* aloe; which is, that it never flowers till it is an hundred years old."[118] Here, Miller is probably referring to the *Aloe americana muricata*, or "the common large American Aloe," which he notes "flower[s]. . . . but once during the Life of the Plant."[119] However, the American aloes are "very hardy . . . [and] are kept in a common Green-house with Oranges, Myrtles, &c" and would have been an odd choice to depict as a rare or special plant.[120] Yet to assume that this is a mistake is to undermine Edgeworth's own substantive botanical knowledge. The often-unreliable nature of scientific texts in this age of discovery and the rapid expansion in botanical knowledge could have misled Edgeworth during her research, but I find another explanation more compelling. Instead, I propose that by imaginatively enmeshing multiple types of aloes together into one that serves both narrative and pedagogical functions, Edgeworth is purposefully creating a fictional botanical hybrid with all of the boundless potential for disruption and creativity that such hybridity enables.

The novel's broader investment in other types of hybridity, while often inconsonant, generally reveals hybridity as a powerful force. While some hybridity is generative, like the blended family of the Percivals and Helena Delacour, other types of hybridity are undesirable, like the creole Mr. Vincent and the threat to white Englishness that his potential marriage to Belinda represents. The aloe, as a colonial object (an acclimatized plant) utilized for domestic purposes (party decor), represents an imperialist hybridity whose influence has tremendous repercussions. The aloe is responsible for both Lady Delacour's injury and her redemption; likewise, its presence in the novel reinscribes her scientific ignorance as it affirms Edgeworth's expertise. Finally, the physical presence of the aloe is echoed by its narrative repercussions: it has the power to move the novel in specific ways toward complication and, eventually, resolution.

Edgeworth also deploys the concept of a fictional hybrid during other scientific developments in the novel. Specifically, she evokes notions of hybridity in the dispersal of Juba's obeah-woman. A multidimensional hybrid figure—an obeah-woman appearing in England, rather than the West Indies, and a religious practice generated by the displacement of enslaved Africans—this ghostly image has the power to temporarily disrupt Juba's peace of mind and Belinda's relationship to Clarence Hervey.[121] When Belinda serves as the voice of scientific reason, exposing the manifestation as merely a "figure . . . drawn with phosphorus," she evinces an intellectual acuity that "first turned Mr. Vincent's attention towards [her]" and incites the beginning of a brief, hopeless courtship.[122] And while Mr. Vincent's attentions and Hervey's putative engagement with Virginia St. Pierre only fleetingly divert Belinda's marriage plot away from its eventual end, the obeah-woman incident reaffirms the ways in which hybrid colonial objects—West Indian religious figures, aloes—have the power to disturb narrative cohesion and, at least temporarily, present alternative possibilities. The obeah-woman injects an array of colonial potentialities into the novel, from interracial marriage to the viability of a non-Judeo-Christian religious perspective. The aloe, on the other hand, reveals the prospect of new avenues for female satisfaction and enlightenment (even if that satisfaction is, occasionally, flawed).

In this respect, the relationship between Edgeworth/Belinda/Delacour and the aloe embodies the kind of special allegiance critics have envisioned between women and botanical life. Edgeworth's decision to tie Lady Delacour's body to the physical presence of the aloe (through its narrative link to the duel with Mrs. Luttridge and the injury to her breast) reinforces her relationship with plants as one of intimacy and interchange. While the intimacy is of a negative cast—the aloe is associated with a violent and competitive relationship that causes what Delacour believes to be breast cancer—Edgeworth ultimately converts that intimacy with the organic world into a positive one, as we see later in the novel.

Moreover, the cultural significance of the aloe, whatever the species, is crucial to its designation as the point of contention between Delacour and Luttridge. Although quite old by the time of *Belinda*'s publication, the botanical volume *Pharmaco-Botanalogia* (1723–27) contains a useful and extensive description of the pre-Linnaean botanical classification, habitat, and potential uses of the aloe. While the author Patrick Blair dismisses the aloe as being "of little or no use in Physick in these Northern Climates," its "juice" (or parenchyma tissue) "is so universally known in the Druggist and Apothecaries Shops" that it was practically commonplace.[123] He further observes casually that "the Gardens of the Curious have of late Years been so well stock'd with a great variety of its Species."[124] Acclimatization gardens clearly had aloe quite frequently in the early part of the eighteenth century in hothouses, as demonstrated via Miller. Artists were also frequently intrigued by its prickly leaves and brightly colored blossoms: Georg Dionysius Ehret produced paintings of multiple different species of aloe throughout his career, it appeared periodically in William Curtis's *Botanical Magazine,* and Pierre-Joseph Redouté included an aloe in his *Histoire Naturelle des Plantes Grasses* (1790). Edgeworth could have seen any of these prints or engravings if she had any interest in botanical artwork or read the *Botanical Magazine* with any regularity. The aloe was clearly a massively popular plant that inspired a significant body of creative work—including portions of this novel. As a plant with healing properties used for unfriendly purposes, the aloe also reflects the tension between Delacour's negative impulses and her genuinely good inner self. By creating a parallel between plant body and human body, Edgeworth cannily takes advantage of the metaphorical relationship between women and plants to envision a more nuanced association. Rather than comparing Delacour to a wilted flower (or even a monstrous plant of the kind envisioned in Darwin's *Botanic Garden*), Edgeworth imagines a connection between an ugly aloe plant—seemingly bad on the outside, but actually wholesome and useful on the inside—and a woman in need of redemption whose very character echoes that same division.

As far as decorative plants go, the aloe is an odd choice: it is not particularly pretty, nor can it stand on its own as an inspiring plant. Delacour reports, "The aloe, you know, is of cumbersome height for a supper ornament. My saloon luckily has a dome, and under the dome we placed it. Round the huge china vase in which it was planted, we placed the most beautiful, or rather the most expensive, hot house plants we could procure. After all, the aloe was an ugly thing—but it answered my purpose."[125] No expense has been spared in an effort to outshine her rival

Mrs. Luttridge, even to the point of purchasing "the most expensive hot house plants," regardless of their botanical or even aesthetic value. The slippage between "most beautiful" and "most expensive," while obviously attesting to Lady Delacour's defective and superficial perspective, also indicates a marked lack of botanical awareness. Delacour can neither tell nor does she care whether a plant is beautiful (the plant on which she spent fifty guineas, after all, is merely "an ugly thing"), only that it serves as a display of her wealth and privilege. Just like the young girl who "runs through a course of natural history, hears something about chemistry, [and] has been taught something of botany," Delacour's intellectual display is flawed and incomplete.

While the partygoers are just as fashionable as she and therefore do not condemn her botanical showiness as vulgar, Belinda—as listener and moderator—hears and judges the unfortunate Delacour. Following the full recital of Delacour's "history," Belinda cycles through a variety of emotions: "Astonishment at her inconsistency—pity for her misfortunes—admiration of her talents—and contempt for her conduct."[126] And, indeed, Delacour's conduct is quite contemptible: her actions in this history do not begin with her flamboyant display of partial botanical knowledge, but, rather, with an ill-conceived and unladylike duel between herself and Mrs. Luttridge, in which a backfiring pistol produces an abscessed bruise that Delacour later becomes convinced is cancerous.[127] This pseudocancerous wound continues to plague Delacour until near the conclusion of the novel, when, through Belinda's care and Dr. X—'s intervention, she is disabused of the notion that she is dying. While many scholars have written about the significance of Delacour's bruised breast, I want to point to the connection between her injury and her misuse of botanical knowledge.[128] As part of the same feud between herself and Mrs. Luttridge, she makes both ethical and botanical mistakes, erring in social behavior and scientific knowledge in the same history. The errors may be committed in tandem, but so, too, are the reparations.

The rehabilitation of Delacour revisits both of these incidents at a crucial moment in the text: just as she is awaiting her mastectomy and mere days before Dr. X— pronounces her wound noncancerous. At this point, Delacour herself is convinced that she will die from the procedure (and is already dying from the cancer), the household is in hectic disarray, and even the usually apathetic Lord Delacour is deeply distressed at his wife's pain and suffering. While her husband attempts to gift her with a diamond ring, Lady Delacour replies that she would prefer a different sort of deathbed present—an act to compensate for her manipulation of the poor gardener and his aloe. Edgeworth summarizes, "Her ladyship then expressed her wish, that a small annuity might be settled upon a poor old man, whom she said she had unwittingly injured. She told the story of the rival galas and the aloe."[129] By making what Delacour believes to be her deathbed confession about the aloe, Edgeworth is reminding the reader of the dangers of partial scientific knowledge. Just as imperfect botanical knowledge led to Belinda's conflicted opinion of her host, imperfect medical knowledge (continually reinforced by a quack doctor) has led to Delacour's purportedly fatal condition and her very real decline from a narcotics habit.

As with many of the scenes in the novel, reparations here come about because of Belinda's actions. Helena Delacour has purchased plants from the very same gardener, whom Belinda has subsequently invited to the Delacours' home—both to deliver the plants and to remind Lady Delacour of her role in changing his fortune. (Recall that Belinda knows the truth behind the aloe from Delacour's earlier confession.) The poor gardener had in fact lost either his employment or his financial security as a result of the disadvantage he was in because of the sale of the aloe to Lady Delacour and was saved from poverty only by the generosity of Lady Anne Percival.[130] Here is more evidence of the eighteenth-century novel's tendency toward fortuitous encounters, another way in which the novel form itself enables Edgeworth's delineation of botanical reparations. By narratively circling back around to the gardener and his aloe, Edgeworth imagines a world in which botanical transgressions can be vicious and cruel, just like many other, more quotidian sins. More important, however, such transgressions are forgivable sins. Delacour has nearly ruined a man's life over a plant, but she is fortunately given the opportunity to compensate for her actions.

This juncture in the novel also highlights the carefully orchestrated performance of Belinda herself. By purchasing her plants from the tricked and betrayed gardener at Belinda's direction, Helena becomes the unwitting conduit for exposing the gardener's sad state to Lady Delacour. As a result, Delacour's botanical and social failures return to trouble her at what she believes is the moment of her death. Yet, through Belinda's moderating actions, Delacour is afforded the chance to make amends for both mistakes. Delacour learns that the Percivals had until recently employed the gardener at an estate in Twickenham, when they had been obliged to leave the house and part with all the employees. She responds both by asking her husband for that small annuity and requesting that he rent the house at Twickenham to allow the gardener to resume his employment.

By acknowledging the gardener's emotions and life experiences, Delacour regains some of her goodness; by respecting his botanical knowledge, she is finally admitting her own flawed scientific knowledge and the problems it has caused. The narrator reports, "The house was taken, and the old man hired as a gardener—a circumstance which seemed to give him as much pleasure as the annuity; for there was a morello cherry-tree in the garden which had succeeded the aloe in his affection: it would have grieved him sorely, he said, to leave his favourite tree to strangers, after all the pains he had been at in *netting* it to keep off the birds."[131] In this happy vignette, botanical knowledge once again plays a significant role. The gardener's affection has been transferred to a new plant and his vocation restored out of what was otherwise sure to be an impoverished end to his life.

Morello is a sour cultivar of acid cherries (*Prunus cerasus*), which were popular in Britain as a food item, but could also be used as decor for their beautiful, bountiful blossoms. Unlike the ugly aloe, the morello has blooms robust enough to serve as centerpiece of a country garden and a pleasant scent to perfume the air. Likewise, as a fruit-bearing tree, it had practical value for an economical household of the kind that Lady Delacour ought to be running. It also required as much care and devotion as the aloe did, including pruning, soil monitoring, and consistently providing proper moisture, allowing the gardener to exercise his craft and find pleasure in his work. Given this botanical outcome, my earlier question of the gardener's ignorance about the value of the aloe seems answered: this is a man of expert knowledge and upright morals whose love for the aloe was perhaps a bit overzealous, but not purposefully deceptive or utterly ignorant.

As a replacement darling, the morello cherry tree would have been an easy-to-find plant. The morello was native to the Levant and temperate zones of Asia, but had been acclimatized to Britain for many generations and was included in *Species Plantarum*—a typical indicator of being a well-established species.[132] It also appeared in many gardening and botany books in England at the end of the eighteenth century, including William Hanbury's *A Complete Body of Planting and Gardening* (1770–73), Samuel Fullmer's *The Young Gardener's Best Companion* (1786), William Mayrick's *The New Family Herbal* (1789), and John Hull's *Elements of Botany* (1800). Even if Edgeworth had never seen such a tree herself, its existence was clearly common knowledge. It was a frankly unremarkable tree, then, without any of the purported exoticism of the aloe, and certainly without the supposed financial value (although, as Edgeworth speculates in "Forester," the cherries themselves could be sold for a profit).[133] Both Delacour and the gardener have learned their lesson to appreciate useful plants over showy exotics, and Delacour has discovered that the people best equipped to appreciate plants are those who fully embrace a robust vegetal knowledge, whether as gardener or botanist. Lady Delacour will never acquire the level of naturalist knowledge possessed by Lady Percival, Belinda, or her own daughter, but her appreciation of its value is enough for her future role as a good wife and mother. Lady Percival and Belinda have already shaped Helena's education; all Lady Delacour has to do is appreciate and support her daughter and friends.

Ultimately, botany serves a healing function in more ways than one—the nameless gardener gets his livelihood back, and Delacour gains the strength through her good deeds to delay the mastectomy. In delaying the procedure, she wins enough time for Belinda to convince her to allow Dr. X— to examine her, thereby heralding the end of her illness.[134] Through her abandonment of the fantasy of supremacy-via-aloe, Delacour finally relinquishes her rivalry with Mrs. Luttridge, among other petty social concerns that had prevented her from assuming her proper role as a domestic figure. And last, but certainly not of the least significance, in encouraging and enabling Helena to give her mother a special plant, Belinda also triggers the final stages of their reconciliation. As with the circulation of natural history knowledge among the Percivals, Helen, and Lady Delacour, Belinda is the mediator for appropriate botanical knowledge and the means by which botany becomes a healing force for multiple women in the novel. In painting Belinda as the centerpiece of such female networks of scientific knowledge, Edgeworth emphasizes botany's significance as a collaborative science. Further, by placing a plant at the center of a crucial narrative moment, Edgeworth affirms the idea that botany can and does hold cultural significance beyond its importance as a genteel field of study for bored or otherwise unfulfilled women.

Belinda is not the only woman whose role in the novel is shaped by creative depictions of botany; Helena Delacour's active scientific pursuits also encompass significant interaction with botanical subjects. I have already established that her plants have narrative value: by purchasing the plants from the old gardener, she incites the beginning of her mother's healing process. But what about the botanical value of the plant in question? While a visitor to the flower sale at Mrs. Dumont's, Helena has purchased a "white Provence rose," or what could be *Rosa centifolia*.[135] This hybrid plant was produced by Dutch horticulturists beginning in the seventeenth century.[136] It is listed in *Species Plantarum* and had been cultivated in England (exclusively in professional gardens) since at least the time of John Ray, who mentions it in his *Historia Plantarum* (1686–1704).[137] It was varyingly known as the Provence rose, or cabbage rose, for its profusion of petals—"centifolia" meaning hundreds of petals. The *Rosa centifolia* also appears with a long list of varietals in Lady Charlotte Murray's *British Garden* (1799), although the term "Provence" does not appear on the list. It was, however, valuable as a medicinal plant (and is still in use today, typically in essential oils). In purchasing the *Rosa centifolia* and giving it to her mother, Helena represents the botanical powers of healing and all the positive valences of scientific study—namely, its ability to create meaningful intellectual interactions for women and its facility in teaching women to be better people.

Even more significant for this passage is what, exactly, *Rosa centifolia* accomplishes. It is listed in William Cullen's *A Treatise of the Materia Medica* (1789) as a purgative and John Aikin's *A Manual of Materia Medica* (1785) as a laxative.[138] Both volumes concur that the rose itself can be made into a syrup, which, when administered to an ill patient, has the effect of cleansing the body of toxins. From a narrative standpoint, the gift could hardly be more appropriate. Lady Delacour has been suffering the side effects of heavy narcotics usage (apparently over a number of years), and a purgative would be an effective way to remove some of the lingering chemicals from her body. The metaphorical valences are striking as well. In learning to be a good mother and wife again, Delacour must figuratively "purge" herself of past coquettish behaviors and inappropriate thoughts (including those incited by her "methodistical" reading).[139] Moreover, on the road to recovery from her breast abscess, she must also emotionally and spiritually purge herself of past wrongs, including compensating the poor old gardener for the loss of the precious aloe.

But I must admit that the white Provence rose is also a beautiful flower. Purchased at a flower market by a daughter eager to resume a relationship with her estranged mother, it may represent to the Delacour women nothing more than a pretty blossom, an offering of peace and love. As a white bloom, it might also represent simply Helena's innocence and virtue, in contrast to her now-almost-redeemed mother, whose past behaviors have led to embarrassment, injury, and even the death of her erstwhile beau Colonel Lawless.[140] However, given Edgeworth's scientific knowledge and her investment in promoting the salutary benefits of naturalist study in *Belinda*, it is more likely that the beauty and innocence of the Provence rose is coupled in the text with its beneficial medicinal properties. This flower is a gesture of peace and healing in multiple vectors, a metaphor for love and purity and an actual purgative to purify the body. Helena thereby represents the generative type of botanical knowledge that her mother failed to acquire through a life of fashionable reading, as well as the useful type of natural history knowledge that her mother failed to teach her. In this, Helena becomes the promise of the future and the hope for Lady Delacour's redemption, as mother and daughter are reunited in both domestic and scientific modes. With her naturalist knowledge shaped by Lady Percival and Belinda, Helena also epitomizes the value of female scientific networks, and the potential of such networks to be enlightening and radically rewarding for those involved.

Building Networks of Scientific Knowledge

Edgeworth's investment in promoting scientific knowledge for women is evident throughout a variety of her publications, from *Letters for Literary Ladies* to *Practical Education* to *Belinda*. In the last, such scientific advocacy relies in part upon the imaginative (and occasionally unlikely) scientific scenarios that punctuate the novel, thereby taking advantage of prose fiction as a unique form by which to convey the difficulties of navigating scientific knowledge as a woman. Unlike the more didactic forms of conduct literature or educational manuals, the novel as a genre allowed Edgeworth to more fully explore the flexible creative possibilities that natural history offered to women in a period where women's paths in life were inevitably restrictive. Further, by calling *Belinda* a "moral tale" rather than a novel, Edgeworth is also evading accusations like those she leveled in *Letters for Literary Ladies*, in which fictive forms of writing could lead to dissipation. Instead, this moral tale leads only to edification and improvement.

Above all, Edgeworth's pedagogical philosophy emerges as one in which women's intellectual networks hold a special significance. One of the most important points of focus in *Belinda* is the appropriate management and transmission of scientific knowledge, in which Belinda herself stands in as moderator for multiple types of scientific information, both in natural history and in botany. Other female characters in the text varyingly represent flawed or idealized scientific knowledge, including Lady Delacour, her daughter Helena, and Lady Anne Percival. Most significantly, the development of scientific knowledge is linked in crucial ways to important narrative moments in the text, including (but not limited to) Lady Delacour's recovery from her illness, her reconciliation with her daughter, and her righting of past wrongs.

Given Edgeworth's early and frequent endorsement of "that habit of industry . . . the love of knowledge and the power of reasoning," it is easy to see why *Belinda* is traditionally interpreted as a case study for various kinds of domestic education. What I have been arguing, however, is that her interest extends beyond conventional literary education, and even beyond the emphasis on education for the sake of domestic stability. In *Belinda,* Edgeworth posits scientific education as a valuable resource that can not only provide happiness and contentment, but that also has the potential to enact meaningful change in a woman's life. And although she repeatedly reminds her readers that she does not intend to turn women into botanists, mathematicians, or chemists, she has provided in her most popular novel a world in which knowledge of science can be invaluable. My reading of the novel also emphasizes the fact that Belinda's moderation of scientific knowledge is directed largely toward healing, pleasing, and helping women. We can and should read Lady Delacour's rehabilitation and Belinda's own future marriage as evidence of the enduring power of the patriarchal system, but it is vital to acknowledge the networks of female knowledge that permeate the entire novel.

Further, as Edgeworth's correspondence proves, such networks of intellectual exchange among women were significant not only in her fiction, but also in her own life. Ultimately, this suggests to me that Edgeworth was invested in the possibility of women participating in multiple networks of knowledge (both domestic and scientific) that are capable of not only coexisting but also of flourishing within the admittedly narrow scope of an eighteenth-century woman's life. The significance of such networks to a broader history of women and science cannot be underestimated; the strength of women's scientific networks in the eighteenth century relied upon the kind of community-building potential espoused by Edgeworth across her prose fictions and pedagogical works. Just as Haywood imagined scientific practice undertaken by groups of adventuring ladies and Priscilla Wakefield proposed the exchange of botanical letters as a means of sisterly bonding, Edgeworth, too, envisioned a world of scientific practice in which women relied upon other women in good times and in bad, sharing knowledge, support, and joy.

Chapter 4 begins the second section of *Botanical Entanglements,* in which I turn from didactic texts used to frame women's scientific authority within the context of their roles as educators and mothers to a selection of artistic productions through which women seized the opportunity to reimagine their relationships with individual biota and the natural world as a whole. Sharing certain elements in common with pedagogical texts, the two projects studied in the next chapter—Elizabeth Blackwell's *A Curious Herbal* and Henrietta Maria Moriarty's *Viridarium*—combine an educational impulse with an advocacy for a different approach to naturalist artistry as well as new ways of relating to plants. The inclusion of the visual arts in a study of women's naturalist labor in the eighteenth century not only expands the parameters of their participation in scientific discourses but also, and more significantly, underscores the blossoming of female-plant intimacies throughout the century.

4

Sketching Vegetality

Blackwell, Moriarty, and Illustrated Botanical Texts

> I have always thought it possible to be a very great botanist without knowing so much as one plant by name; and without wishing to make your daughter a very great botanist, I think nevertheless that it will always be useful to her to learn how to see, whatever she looks at, well.
>
> —Jean-Jacques Rousseau, *Lettres élémentaires sur la botanique*

While text was the primary medium through which women in the eighteenth century intervened in naturalist discourses, artistic productions were equally significant to the elucidation of women's scientific labor in the period. I look to Elizabeth Blackwell's *A Curious Herbal* (1737–39) and Henrietta Maria Moriarty's *Viridarium* (1806) as two examples of illustrated botanical works that merge the pedagogical qualities of the texts studied in the previous three chapters with the kind of radical creativity examined in this chapter and the subsequent sections of *Botanical Entanglements*. Illustrated texts that present a variety of plants both native and exotic, these two volumes represent some of the most significant contributions by women to the history of botanical illustration in England. Blackwell and Moriarty offer not only botanical projects to which they contributed at nearly every stage of production—a rarity for women in the period—but also a distinctive approach in those projects to conceptualizing botanical authority and vegetal autonomy. In what follows, I argue that Blackwell and Moriarty developed their own methods for working in the botanical sciences by combining established techniques and theories with a creative relationship to individual plant specimens. Both author-illustrators also develop in their reference works what I refer to as a "female botanical subjectivity": each work creates an immersive, pedagogical experience for its women readers by inviting them to envision themselves in a specific garden as part of the process of learning about botany. Thus, instead of adopting a position of authority over the natural world, as did so many professional artists and botanizers, these women worked alongside and within it. With nature typically gendered feminine in the eighteenth century, such metaphorical connections between women and their spaces of botanical labor enabled the assumption of scientific authority, as I have shown in previous chapters.

In some ways, this chapter's epigraph from Jean-Jacques Rousseau typifies the cultural climate within which these and many other women undertook naturalist labor in the eighteenth century. In fact, Rousseau's advice to his friend Madame Étienne Delessert—part of a series of letters designed to teach her young daughter Marguerite-Madeleine the basics of botanical study—was among the most circulated botanical texts of the late eighteenth century. Translated into English and significantly expanded by Thomas Martyn, a professor of botany at Cambridge University and a Fellow of the Royal Society, *Letters on the Elements of Botany* was popular with women naturalists and educators alike, but it presents a misleading portrait of women's involvement in natural history in the eighteenth century. In spite of the fact that he was expressly promoting scientific study for young girls by virtue of the publication's intended audience, Rousseau's stance on botanizing women was, in effect, rather conservative. In the excerpt above, his language indicates not merely a resistance to the prevalence of Linnaean taxonomy or a rejection of its relevance to young girls, but also, and more important, a demand for that excruciatingly fragile balance in women's scientific labors with respect to how much scientific knowledge women could possess. According to Rousseau, a woman should not wish to be "a very great botanist"—an attitude seemingly consonant with the kind of careful hedging that Maria Jacson would practice later in her *Botanical Dialogues* (1797)—nor is it any good for her to learn "so much as one plant by name" (which then abruptly diverges from Jacson). Rather, botany is valuable for the fact that it teaches girls "how to see . . . well" and how to utilize a connection with nature to make them better people—and, later, better wives and mothers. Botanical study for women could never be an end in itself or a source of sheer pleasure, but should instead serve a larger aim in developing young ladies into moral and responsible adults.

However, women often objected—actively—to this diluted version of scientific study put forth for their consideration. Although striking a balance between active intellectual labor and appropriate femininity was a paramount concern to many of the naturalists of this project, the work these women produced also exceeds the boundaries imagined by conservative commentators such as Rousseau. Illustrated botanical reference works were immensely popular in the eighteenth century, and women artists played a significant role in cataloging and documenting both native and acclimatized species. Together, Blackwell and Moriarty depicted hundreds of species of plants in their illustrated texts. While this represented but a fraction of the biota being grown in botanical gardens and greenhouses during the period, their choice of these particular plants does indicate to a modern audience which plants eighteenth-century naturalists thought the most significant or notable. Within their respective illustrated texts, however, both Blackwell and Moriarty presented new possibilities for studying individual specimens as part of a broader selection of indigenous and acclimatized plants. I argue that by paying attention to not only the precise selection of plant subjects but also the manner of their depiction, we see how Blackwell and Moriarty's approach to botanical illustration evokes the allegiance with organic plant life identified by modern critical theorists.

It is in this unique approach to the presentation of botanical information that I trace their formation of an expressly female botanical subjectivity. In some ways opposed to, while in others akin to, traditional approaches to botanical illustration and study, this female botanical subjectivity importantly differs from the conventional objective approach to botanical labor and botanical standards of production used by male naturalists. Where men like Georg Dionysius Ehret followed a very particular set of standardized guidelines regarding layout, color, composition, and even inclusive text in their botanical illustrations, Moriarty imprecisely adapted those guidelines for her own critical purposes. Likewise, earlier in the century Blackwell departed from trends in herbal illustration for *A Curious Herbal* to present vivid evocations of real plants in real gardens. To some extent, this difference is rooted in what Stephen Shapin and Simon Schaffer have identified as the gender-based dynamics of "modest witnessing."[1] A deeply influential perspective on the way modern experimentation was supposed to work, this restriction of objectivity to a very specific type of male witness meant that women were not only excluded in practice from scientific spaces like the Royal Society's facilities at Gresham College, but in theory from the capacity to hold any kind of scientific authority whatsoever. Blackwell and Moriarty make no claims for their own ability to be modest witnesses, but they do put forth a distinctive botanical authority based upon their gender through their prefatory materials, layout and design, and artistic approach to the images. Furthermore, their "witnessing" as such is sympathetic and empathetic, recognizing the individuality of plant specimens rather than evaluating each by its degree of adherence to a type specimen. In embracing their knowledge of specific plants in specific gardens, Blackwell and Moriarty lay claim to a new, different kind of botanical authority. By conceptualizing this authority as a purposeful subjective stance that rejects "typical" relationality to nature, we open up space for understanding how these women created a shared botanical experience among their readers for innovative pedagogical purposes.

This female botanical subjectivity was articulated within the framing of the specific gardens from which Blackwell and Moriarty drew their inspiration and specimens. Stephen Bending has traced a tradition of women's involvement in eighteenth-century garden spaces that highlights the profound ways in which many women cultivated significant relationships with specific gardens, emphasizing that "the shaping of physical space is the shaping also of identity."[2] I expand upon his insightful study to examine how Blackwell and Moriarty used their personal connections with individual horticultural spaces not only to participate in broader botanical discourses, but to rearticulate the dynamic between themselves (as individual women, and as representative of women as a larger population) and plants, between human and vegetal life. Both Blackwell and Moriarty based their projects almost entirely on specimens found in a specific garden space, reinforcing their claims to authority by providing a tangible locus for their data. Likewise, both *A Curious Herbal* and *Viridarium* also evoke larger implications about the meaning of gardens to women's scientific praxis. In their efforts to produce original scientific content and assert their own artistic voice, these women utilized the productive metaphorical and physical space of the garden as part of their distinctive approach to botanical artistry.

A Brief History of Botanical Illustration

The history of botanical illustration is a long and fascinating one. From the earliest technical illustrations of plants in the fifteenth century BCE to lushly painted florilegia in the eighteenth century, the style, scope, and intent of botanical illustration changed significantly over time. Botanical illustration as we now think of it began with herbals, which are essentially a record of plants with medicinal or culinary uses. The most famous English early modern herbalists are John Gerard, whose *Herball, Or Generall Historie of Plantes* (1597) was well known throughout England in the eighteenth century, and John Parkinson, whose *Paradisi in Sole Paradisus Terrestris* (1629) and *Theatrum Botanicum* (1640) were recommended for medical students in the seventeenth century and beyond.[3] For its comprehensive and extensive record of medicinal plants, Gerard's *Herball* remained a common point of reference for botanical writers even by the early nineteenth century; Maria Jacson is only one of the many female authors who relied upon his herbal as a point of reference in their writings. In spite of its many errors and the fact that the woodcuts were copied from Continental sources, it was incredibly popular and influenced the production of later herbals, including Elizabeth Blackwell's own *A Curious Herbal*.[4]

The seventeenth and eighteenth centuries saw a great flourishing of botanical artwork when explorers and botanists from Europe and England began to pursue exotic specimens in earnest, frequently supported by royal funding or other commercial investments. Often, botanical research and artwork served imperialist purposes, as was the case even with women's work that circulated in mostly domestic spheres of consumption. Likewise, as an interest in developing a comprehensive yet usable classification system for such imperial projects rose in the seventeenth century, botanical artwork began to adhere to a certain set of conventions. It was first and foremost necessary that an illustration of a plant specimen be recognizable as an example of a given plant; it needed to possess the identifying markers that were deemed important by different classificatory systems, including leaves and petals (pre-Linnaeus) and reproductive parts (post-Linnaeus). It was also important that the image represented an ideal version of the plant, rather than a temporally located one; theoretically, the sketch produced by a talented botanical artist would represent a composite image derived from a number of different specimens instead of a faithful record of an individual plant—an adherence to essentialism rather than realism that led to some tension between naturalists (who were not typically artistically gifted) and their hired botanical artists.[5] Finally, by the eighteenth century, it was also important for a botanical illustration to illuminate the full spectrum of a plant's growth, rather than the present stage alone. So, for example, a fruit-bearing plant would be illustrated with seeds, buds, flowers, and fruit—sometimes on the same plant (a physically impossible specimen) and sometimes in separate images on the same page.

Blackwell's *A Curious Herbal*, for example, in a plate featuring the blackberry bush (plate 45) displays the blossom, the ripe berry, and an unripe berry on the same page but not the same individual plant.[6] Similarly, Moriarty's illustration of the Winged-Podded Sophora (plate 44; *Sophora Tetraptera*) includes the blossoming plant, the leaves, and the "pods," or unopened buds.[7] Later in the century, botanical illustration also incorporated cross-sections of a plant, showcasing reproductive parts both internal and external, as well as other internal anatomy deemed important. As such depictive trends developed, "botanical art in late eighteenth-century Britain was heavily influenced by its role in the classification of plant life, and in effect, became an extension of botany."[8] Art and science became more closely linked, both on and off the page. Enabled by the evolution of less expensive printing technologies and highly motivated by the increased demand for botanical works, the volume of botanical artistry in the period was remarkable.

While botanical artwork evolved alongside classificatory systems, so, too, did the places where plants were cultivated. As seventeenth-century England saw a massive influx in new plants at professional gardens like the newly founded Chelsea Physic Garden (opened in 1673), such changes in the makeup of organic life also ushered in a trend in pleasure gardens and a demand for exotic specimens for private households. Particularly as physical spaces like botanical gardens and pleasure gardens blurred the lines between aesthetic object and scientific subject, it becomes necessary to think about botanical artwork as at once a creative act and an epistemological commitment. As Stephanie Ross has compellingly outlined in *What Gardens Mean* (1998), the physical space of the pleasure garden was often intended as an artistic project that shared certain key qualities with other art forms such as poetry and painting. To extend this logic, paintings done after specimens in or from a botanical garden (which was common among women artists, who rarely had access to foreign specimens in their native contexts) thereby multiply the artistic dimensions of the given image: art imitates art, in a sense. This suggests that we cannot evaluate such a creative genre on a scientific axis alone; eighteenth-century botanical artwork productively and provocatively combines creative and scientific fields. Indeed, as Tita Chico has shown in *The Experimental Imagination* (2018), creative and scientific labors in the eighteenth century were mutually constitutive, with literary arts and their imaginative tropes often enabling the articulation—and therefore production—of scientific knowledge. I propose that Blackwell and Moriarty, along with other women naturalist-artists of the period, were uniquely positioned to take advantage of the intersections between science and art.

Women Botanical Artists

Although many of their creative productions have been lost to time, women in the eighteenth century actively included botanical subjects in a variety of domestic art projects, from the sketches of a flower bouquet on a table made by a child to the meticulous paper mosaics of the seventy-something Mary Delany. Significantly, the production of botanical artwork was also appealing as a creative means by which women could participate in natural history as both artists and botanists. Because of its similarities to flower painting, it was considered an acceptable artistic practice among women of the middling or upper classes. Botanical illustration was thereby seen as both genteel and appropriately challenging for the female intellect, as well as a respectable means of earning some money if necessary.

We see this evidenced in the historical record, as well as in popular fiction of the period. For example, the all-female society in Sarah Scott's *Millenium Hall* (1762) encourages its members and pupils in the arts, here encompassing both drawing and gardening in a combination that surely gave rise to some botanical sketches.[9] Similarly, the talented Adelaide in Elizabeth Cullen Brown's *The Sisters of St. Gothard* (1819) sketches and paints natural scenes as part of her daily recreation.[10] In another context, the impeccably moral Laura Montreville in Mary Brunton's *Self-Control* (1811) supports herself for a time in London by selling sketches and paintings. In a more specific example, Althea Dacres of Charlotte Smith's *Marchmont* (1796) learns to draw flowers by copying botanical plates and then proceeds to draw plants directly from nature—a typical developmental trajectory for most botanical illustrators in the period.[11] Fictional representations of women engaging in artistic endeavors—for money or pleasure—are more common in the early nineteenth century, as may be seen further in novels such as Frances Burney's *The Wanderer* (1814) and Jane Austen's *Emma* (1815). As this brief survey suggests, women were certainly learning artistic skills and employing those skills in depicting various flora throughout the period. Women's participation in the field of eighteenth-century botanical illustration thus combined both recreation and employment.

History tells a longer story. Women were renowned as flower painters in England and Europe from the seventeenth to nineteenth centuries, including the Dutch Rachel Ruysch, the French Anne Vallayer-Coster, and several British artists such as Margaret Meen, Mary Moser, Anne Pratt, and Jane Loudon. Moser was a founding member of the Royal Academy best known for her elegantly detailed oil paintings and a prestigious commission from Queen Charlotte to decorate a room in Frogmore House with her signature florals.[12] Her paintings also reflected a fondness for native English species, unlike Ruysch or Vallayer-Coster, who instead painted elaborately detailed portraits of exotic and native specimens alike.[13] Pratt and Loudon, on the other hand, combined botanical studies with their artwork, illustrating scientific volumes of their own writing. Pratt's *The Flowering Plants of Great Britain* ran to five volumes (with a sixth added much later to incorporate grasses, sedges, and ferns), and Loudon's several gardening manuals were popular for years after their publication.[14] While both of these women published their works in the nineteenth century, they were part of an artistic and intellectual legacy that stretched back to Maria Sibylla Merian in the late seventeenth century.

The linkages between flower painting and botanical illustration facilitated women's participation in this growing scientific field. The propriety of women painting—particularly a conceptually feminine subject such as flowers or nature scenes—was unquestionable, and the transition from painting a bouquet of flowers to painting the anatomical details of a flower, although not a painless one, was facilitated by those metaphorical connections between women and flowers. Women's botanical artwork, from herbals to florilegia to gardening manuals, thus provided key opportunities for women to intervene in critical botanical discourses. However, such works have often been viewed as inferior to those of men, particularly in twentieth-century histories of botanical artwork, because of a perceived lack of scientific rigor in texts by women. This is often in part due to the fact that women regularly had other individuals (usually men) write the content and scientific descriptions of the plants in question and preferred to focus exclusively on the visual components of the text. Such assistance was typical when it came to the scientific names of plants or extensive descriptions of habitat, growth patterns, and care. In cases where women did not ask for in-person assistance in labeling a plant, it was also common to copy botanical names from other scientific reference sources, like Philip Miller's perennially popular *Gardeners Dictionary*. This trend would shift somewhat by the mid-nineteenth century, when Pratt and Loudon sought to produce accessible horticultural texts for women by combining visuals with easy-to-comprehend introductions to gardening. These texts, however, have rarely been considered part of a scientific legacy and are instead thought of as part of the casual Victorian interest in the connections between literature, gardening, and flowers—including the development of the language of flowers.

Women's botanical artwork has, unfortunately, often been dismissed for misogynistic reasons. Certain twentieth-century male critics and art historians, for example, refer to Blackwell's work in trivializing ways, by calling her painstakingly hand-drawn and hand-colored etchings "of mediocre standard only" and "the work of an industrious amateur [that] show no touch of genius."[15] Perhaps unsurprisingly, female critics were more complimentary: Emily Read Cheston in a 1942 article written for the *Herbarist* calls Blackwell's work "accurate," "clear," and "admirable."[16] While the technical aptitude of a woman with no professional instruction in artistic skills might be debated, it seems more likely that such a dismissal is the result of an unwillingness to take women's botanical labor as seriously scientific.

On the other hand, just like the many women whose invisible labor underpins significant amounts of literary and historical scholarship of the twentieth century, women often lent their artistic skills to their husband's (or another man's) botanical works without acknowledgement or credit in the seventeenth and eighteenth centuries.[17] It is hard to estimate the number of women who contributed to the scientific works of the men in their lives but, given enduring patterns in relationships between scientific husbands and wives, brothers and sisters, or fathers and daughters, it is relatively safe to assume that there are far more in the history of scientific artwork than we can ever know. As Virginia Woolf famously suggested, an anonymous author was potentially quite often a female author (or, in this case, an artist). Occasionally, however, we do have a record of women providing illustrations for published botanical texts or other scientific works. Women drew or colored for William Curtis's *Botanical Magazine* from an early period, for example, and are today responsible for a significant number of the images in this long-running magazine. Later, in one notable case, Matilda Smith was the primary artist for the publication and contributed more than 2,300 drawings in the late nineteenth and early twentieth centuries.[18] Women like Augusta Innes Withers and Sarah Ann Drake also illustrated a number of British botanical and horticultural publications in the nineteenth century, including *The Orchidaceae of Mexico and Guatemala* (1837–43).

Of course, women drew botanical subjects as amateurs, too. Even if families could not afford the services of a drawing master for their young daughters, the publication of books that taught flower drawing were popular in the eighteenth and early nineteenth centuries, such as *The Lady's Drawing Book and Compleat Florist* (1755), *The Florist, or an Extensive and Curious Collection of Flowers for the Initiation of Young Ladies Either in Drawing Pictures or in Needlepoint* (1759), *Flower Painting Made Easy* (1766), and James Sowerby's *Easy Introduction to Drawing Flowers According to Nature* (1788). Women even entered the field of drawing manuals with Mary Lawrence's *Sketches of Flowers from Nature* (1801), to give just one example. Texts like these often included instructions for painting or sketching flowers in multiple steps. *A New Treatise in Flower Painting* (1797) outlines a very specific program for teaching young women how to paint their organic subjects: "The flowers intended for this work, are engraved in a peculiar manner; and will be shown in three different states: first, the outline; second, the first tints put on, in order to show in what manner to begin to colour or paint a flower . . . thirdly, the flowers finished."[19] Once again, given the overlap between flower painting and botanical artwork in the eighteenth century, this plethora of instruction manuals indicates the broader public's high level of interest in botanical artwork.

While it is plausible that both Blackwell and Moriarty were trained by professional artists as young women, there is good reason to suppose that both women also looked to resources such as these when formulating their respective masterpieces. Particularly in light of Blackwell's ability to sketch, engrave, and paint the images in her herbal entirely on her own, it is probable that she did consult reference texts of some sort in addition to those she utilized to provide cultivation information. She also likely learned engraving as a young woman from a mentor; engraving was a difficult and specific skill that usually only artists with extensive artisan training like Merian were able to exercise.[20] Considering the number of exotics in Blackwell's volumes in particular, it is also important to notice where she specifically notes having not copied her specimens from life—a few plates of specimens were done from Hendrik Adriaan van Rheede tot Drakenstein's *Hortus Indica Malabaricus* (1678–1693). For example, she labels the Indian Nut (plate 387) and the Indian Apple (plate 388) as being from "the Malabar Garden."[21] She was even specific enough to note the volume and table number, so curious readers with access to this publication could consult the relevant volumes themselves. Similarly, Moriarty may have employed guide books when drawing non-native species as a means of finding a reference point in addition to live specimens;

we know that she consulted Curtis's *Botanical Magazine* at least some of the time.

The types of plants and variety of specimens included in *Viridarium* is comparable to those in Blackwell's *Curious Herbal,* which speaks both to Blackwell's expansive research and Moriarty's limited focus on greenhouse plants in particular. The variety of botanical specimens in England by the turn of the nineteenth century was exponentially greater than that of professional gardens in London in the 1730s. According to one estimate, "During Banks' supervision of Kew (1772–1820) approximately seven thousand new plants arrived in England from around the world," making Kew one of the most comprehensive and wide-ranging gardens in the entire globe.[22] Kew was, however, not necessarily a universal resource. While elite men and women had access to scores upon scores of new specimens from Kew or from individuals like Joseph Banks and Daniel Solander, it was not a botanical garden open to the broader public in the late eighteenth and early nineteenth centuries, when Moriarty was producing her work. Likewise, the Chelsea Physic Garden may have been inaccessible to her for a variety of reasons, not least of which was that it was a professional garden dedicated to medicinal and horticultural research, rather than a pleasure garden of the sort typically open to the public. Without the professional sponsorship Blackwell enjoyed, Moriarty may very well have been barred from visiting this working garden. However, as I will suggest below, she likely had access to a private botanical garden or greenhouse.

In Blackwell's case, some plant specimens were simply unavailable in the Chelsea Physic Garden and therefore not accessible as models for her plates. Unsurprisingly, there were also a significant number of plants that acclimatization gardens were unsuccessful at adapting to British climates, even in controlled hothouse spaces. In many instances, these specimens would have to be documented while abroad, preventing most women from participating in such artistic endeavors. Often, amateur draftsmen would capture a rough likeness of a given plant in a notebook, while more accomplished artists back home in Britain would replicate in careful detail the image of the specimen brought back from the voyage. And, of course, others in Britain (often amateurs themselves, frequently women) recreated images of exotic and domestic flora from other pictures or paintings, creating a kind of layered, secondary level of artistic representation. While the history of botanical illustration as told by those with a traditional mindset has often relegated the former category of artists to mere copyists, I think it evident that an accurate history of women's botanical artistry cannot be told without incorporating such individuals. Particularly with regard to Moriarty's *Viridarium*, it is useful to think about the different ways in which women connected to plant life and how such strategies differed from the opportunities available to male artists in critically compelling ways. Female botanical artists like Blackwell and Moriarty not only diverged from male artists in their compositional approach but also in the level of emotional engagement with their subjects.

Kenneth Nyberg has written of "the tension between curiosity and utility as driving forces behind botanical exploration," a paradigm I also want to apply to the field of botanical illustration.[23] Especially for women like Blackwell and Moriarty, financial considerations were a significant factor in the production of their works, as I will elaborate upon below. As those familiar with eighteenth-century history will know, many women who published and sold work in the period did so in part out of financial necessity, including Eliza Haywood, Charlotte Lennox, Maria Jacson, and Priscilla Wakefield. Thus, botanical artwork like that found in *A Curious Herbal* and *Viridarium* may not be "purely" borne of artistic inspiration, as some critics seem to imagine Georg Ehret's work to have been (he also painted in part for financial reasons). Both Blackwell and Moriarty likely had artistic and scientific motivations for producing their texts as well, a possibility that has often been overlooked in previous criticism. I find it highly unlikely that either woman had absolutely no interest in botany or the natural world (an accusation leveled at Moriarty); it is far more plausible that both had a preexisting interest in the science and at least some investment in expanding their botanical knowledge. It is also entirely possible that these women both desperately needed money and enjoyed botanical study for its own sake. Embracing this tension between artistry and pragmatism in the case of these two naturalists allows us to open up space for a more generous reading of women's scientific involvement in the history of eighteenth-century botany. In doing so, we can more fully appreciate the extent to which Blackwell and Moriarty produced compelling work while largely adhering to prescribed artistic techniques and conventions of botanical illustration.

A Curious Herbal for a Scurrilous Husband

Elizabeth Blackwell is arguably the most famous female botanical illustrator of the eighteenth century, having produced one of the most popular and comprehensive herbals in the entire period. Born to the wealthy

Blachrie family in Aberdeen, Scotland, Elizabeth was educated in the arts and modern languages, skills that would later aid her in the production of her famous work. She married her second cousin Alexander Blackwell, an erstwhile physician who, when they moved to London, attempted to break into the printing business. He was soon in trouble with the printer's guild for his lack of training or affiliation and was eventually jailed in Highgate Prison for debt. Elizabeth's motivation for *A Curious Herbal* has been typically attributed to her desire to free her husband from debtor's prison. Bruce Madge has proposed that Blackwell saw a copy of Merian's popular *Neues Blumenbuch* (1680), which may have influenced her desire to produce a botanical work of her own in search of a profit.[24] Botanical reference books were immensely popular in the period and would have been in high demand.

Blackwell completed some preliminary botanical drawings and sent them to Sir Hans Sloane and Dr. Richard Mead (a physician known for his advocacy in favor of Lady Mary Wortley Montagu's inoculation campaign). She would eventually dedicate the first volume to Mead, which suggests that he was especially supportive of her labors. Blackwell took a lodging on Swan Walk near the Chelsea Physic Garden in 1736 and spent the next few years visiting the Garden to sketch five hundred plant specimens almost entirely from live growth. Indeed, it was her proximity to the garden and her access to fresh specimens whenever she needed them that set her herbal apart from many of her predecessors, who rarely worked from live plants. Perhaps predictably, Blackwell disavowed any substantive botanical knowledge in her published work and evoked the typical modesty topos, likely to deflect criticism of a woman publishing, and a woman working in the sciences at that. In her dedication to the second volume, she thanks Royal Society member Isaac Rand for his assistance, declaring: "Your READINESS to ASSIST and INSTRUCT me in this UNDERTAKING (which otherways would have been very imperfect) is so visible to every JUDGE who must know that I had no SKILL in BOTANY that I am under a Necessity to declare THAT IT IS TO YOU I AM OBLIGED FOR EVERY COMPLEAT PART OF THIS WORK."[25] To what extant Rand actually assisted with her production is unclear, but we do know that he continued to support her throughout her endeavors in both intellectual and financial ways.

Blackwell earned the support of several famous botanists, physicians, and gardeners. On the dedicatory page of *A Curious Herbal* she wrote, "The UNDERTAKING was honoured with the following Publick RECOMMENDATION by the Underwritten Gentleman," before listing such individuals as the gardener-cum-botanist Joseph Miller as well as a number of physicians including Alexander Stuart, James Sherard, and the aforementioned Mead.[26] The inscription above the names makes explicit their approval of her artistic and botanical abilities, and the work as a whole: "We whose Names are underwritten, having seen a considerable Number of the DRAWINGS from which the PLATES are to be Engraved, and likewise some of the COLOUR'D Plants, think it a Justice done the PUBLICK to declare our Satisfaction with them, and our good Opinion of the Capacity of the Undertaker."[27] Blackwell's endeavors were thereby supported by prestigious naturalists and intellectuals from the start, suggesting that she enjoyed a privileged position that few other women would have. The first volume of *A Curious Herbal* was also advertised in the July 1737 issue of the *Gentleman's Magazine*, noted there as "printed for Samuel Harding in St. Martin's Lane" and attributed solely to "Mrs. Blackwell."[28] Its placement in a magazine at least ostensibly geared toward a male audience also suggests that her work possessed an intellectual status that elevated her above a simple flower painter.

The first volume was so universally approved of that the College of Physicians provided a testimonial (combined with a "handsome" financial award) calling her work "most useful" and promoting it to the general public.[29] In his 1790 *Progress of Botany*, Dr. Richard Pulteney, a respected physician and botanist who actively promoted the Linnaean system of classification in England, declared that "physic is indebted for the most complete set of figures of the medicinal plants, to the genius and industry of a lady, exerted on an occasion that redounded highly to her praise."[30] The individual issues of four plates with descriptions were released weekly from 1737 to 1739 over the course of 125 weeks, and bound editions soon followed.[31] The reputation of *A Curious Herbal* "remained high to the end of the century, in countries abroad as well as on its own ground."[32] There were several official copies in circulation in the mid-eighteenth century, multiple pirated editions, and an expansive and expensive enlarged edition published later as the *Herbarium Blackwellianum*.

The *Herbarium Blackwellianum* was edited by Christoph Jacob Trew (Ehret's patron and a member of the Royal Society), who added Latin text in place of the English and included supplementary materials such as copious explanatory footnotes. The artist, Nicholas Friedrich Eisenberger, redrew and engraved Blackwell's original plates as well as added 115 plates to this new edition.[33] The *Herbarium Blackwellianum* was published in five volumes from 1750 to 1760, complete with luxe bindings and elaborate frontispieces, making for a far grander production than Blackwell's own work. According to one source, it was also published again in 1794 at Leipzig.[34] Whether Trew intended merely to capitalize on Blackwell's success or had a genuine interest in her botanical endeavor is unclear, but his investment in her work and his "remastering" of her text ensured that her name would remain known in botanical circles long after her death.

In spite of Blackwell's success with her herbal, however, her story does not end happily. While her earnings from *A Curious Herbal* were sufficient to repay her husband's debts, he later bounced around between employment and patronage, going from director of improvements under the Duke of Chandos in Middlesex (a position from which he was purportedly dismissed) to physician to King Frederick of Sweden (at which point he was again accused of quackery) to agricultural author and experimental farm manager (another mismanaged affair).[35] Eventually, Alexander Blackwell became involved in a plot to alter the Swedish succession, and he was executed on August 9, 1747.[36] We know little of Elizabeth's life after his death; she produced no more artistic works (botanical or otherwise) and died in 1758.

A Botanical Legacy

As a singular intellectual and artistic effort, *A Curious Herbal* is certainly quite the legacy. With its detailed descriptions and five hundred painstakingly illustrated plates, the herbal is attractive, useful, and influential. The frontispiece of the first volume, although not likely drawn by Blackwell herself, is a significant component of the herbal's botanical authority. The image shows two seated individuals, identified as Theophrastus and Dioscorides—arguably the two most important figures in classical botany. Dioscorides was responsible for the landmark *De Materia Medica,* which influenced botanical artwork for centuries after its original publication; Theophrastus, while admittedly a polymath, was well known as an early Greek theorizer of botanical science, with his most famous botanical publications being *Enquiry into Plants* (also known by the later Latin name of *Historia Plantarum*) and *On the Causes of Plants,* both of which were written in the first century BCE.

Theophrastus's two works are generally considered the first attempt at botanical systematization and were significant texts in the history of botany even in the eighteenth century: Linnaeus read (and was likely influenced by) *Enquiry into Plants* as a young man.[37] Setting a precedent for generations of herbals, this text organized plants by a variety of subcategories including size, scent, taste, growth patterns, growing location, and their practical uses in medicine and cooking (categories of classification that Linnaeus sought to disrupt with his sexual systematics, but that have enthusiastically reemerged in the age of critical plant studies). Blackwell's *A Curious Herbal,* on the contrary, does not identify a system with regards to the groupings of plants she produced, with periodic exceptions like plates 334–36, which included three different types of fungi in the same batch of engravings. Likewise, Blackwell omitted some of the more fantastical plants often found in replicas or imitations of *De Materia Medica,* such as the bird-producing goose barnacle tree, although she does include the eighteenth century's favorite pseudozoophyte, the Scythian Lamb or Barometz (plate 360). It seems likely, then, that these two individuals in the frontispiece are designed to convey botanical authority and establish an intellectual legacy that *A Curious Herbal* strove to be a part of, rather than to represent any kind of concrete influence on Blackwell's work.

Blackwell's *Herbal* is also significant within the history of eighteenth-century botanical illustration for the fact that it was produced entirely before the Linnaean classification system was introduced to western botanists. Linnaeus's *Species Plantarum* was published in May 1753, more than fifteen years after the commencement of Blackwell's work in 1736. And while Linnaeus had published multiple scientific and botanical works during Blackwell's career, including *Systema Naturae* (1735) and *Genera Plantarum* (1737), they had yet to be translated into English and were not popular in Great Britain.[38] Further, his system of binomial nomenclature—which is about as much Linnaean data as would have been useful to Blackwell—is usually dated to *Species Plantarum,* which was written between 1746 and 1748 and therefore postdates Blackwell's work.[39] It is possible that Blackwell's work was so well received by her male successors and later generations of readers in part because her work was not shaped by the sexual system and thereby avoided any hints of impropriety—a challenge Moriarty faced at the end of the century.

With the Linnaean binomial yet to be codified, Blackwell was producing *A Curious Herbal* in the midst of a number of competing classification systems that would fail to sort themselves out in Britain. The most prominent of these was that of the seventeenth-century English naturalist John Ray, although the classification systems of Joseph Pitton de Tournefort and others also held sway in various parts of Europe. *A Curious Herbal* seems to use Ray's classification system from *Historia*

Plantarum (1685–1703), which was certainly less unwieldy than the previous early modern tradition of describing plants by a seemingly endless series of descriptors. In a practical sense, this rendered *A Curious Herbal* a more accessible text than some other herbals in the period, especially given Blackwell's use of English for all descriptive passages (something that Trew needlessly changed in his adaptation). Blackwell's investment in the pedagogical potential of her text is clear: by avoiding Latin, keeping her descriptions simple and user-friendly, and including highly detailed full-color plates, her herbal echoes many of the basic design principles behind good educational works. Her anticipated audience was therefore not only expert men who might consult the volume for its breadth of species, but an interested lay population that included women.

A Curious Herbal is both curious and useful. The layout of the volume is easy to follow and was likely accessible to a spectrum of readers, whether an experienced midwife or a student chemist. A descriptive page featuring information about four distinct (and sometimes related) plants is followed by four different, full-page images representing each plant. The individual plates featured the pre-Linnaean taxonomic term for the plant, the name of the plant in English, and a key to the illustration itself, which usually consisted of a combination of flower, fruit, seed, and root. The prefatory page was entirely text and included four key subsections: details that help identify the plant such as leaf shape, coloration, and height; the season in which the plant is cultivated or blooms; the medicinal uses of the plant, which sometimes included specific instructions for how to utilize the plant in a recipe; and a series of other names for the plant in different languages, encompassing a wide selection of ancient and modern languages such as Greek, Latin, Spanish, Italian, French, German, and Dutch. Blackwell explains her rationale behind this layout: "The Undertaker, being desirous to make this Work more useful to such as are not familiar with other Herbals, is resolved (for their Sake) to give a short Description of each Plant, the Place of Growth, and Time of Flowering with its common uses in Physick, chiefly extracted from M^{r}. Joseph Miller's Botanicum Officinale with his Consent, and the ordinary Names of the Plant in different Languages."[40] Alexander Blackwell is said to have provided the translated names of the plants from his jail cell, although, as indicated above, Blackwell herself consulted Miller's *Botanicum Officinale* (1722) to provide some of the additional material.[41] Moreover, as historians including Elaine Leong have shown, it was common for women to keep exhaustive records of botanical information in recipe books for both culinary and medicinal purposes—books that were often passed down from generation to generation.[42] It is therefore equally likely that some of the data in *A Curious Herbal* concerning the uses and preparations of certain plants was derived from Blackwell's own family lore, representing a legacy of domestic scientific inquiry and experimentation that was so vital to female practitioners in the eighteenth century.

Plate 55, for example, "March Violet," includes a series of notes about the medical deployments of violets in a cordial or syrup. Used to treat a wide variety of ailments from fevers and inflammations to digestive issues and coughs, violet syrup was a well-known remedy in domestic medicine that most efficient and self-sustaining households in the period would have had a reliable recipe for.[43] Blackwell also indicates that it was specifically used to treat children, likely because of its high sugar content. The specificity and level of detail in this segment of each entry varies significantly, perhaps indicating a division between plants about which she had personal or familial knowledge and those plants (usually exotics) for which she had only technical reference books to supply the information. External resources are regularly if not consistently mentioned as sources for her medicinal remedies (including Parkinson, Dioscorides, and John Prevotius), and Miller's *Botanicum Officinale* is cited explicitly and in detail when used. This combination of knowledge from professional, published texts and information from domestic, communal resources reflects Blackwell's dedication to producing an herbal that would be invaluable to a broad audience and as thoroughly researched as possible—perhaps another one of the many reasons her text was held in such high regard for so long after its initial publication.

Blackwell's selection of plants is also significant most particularly for its reflection of a truly global eighteenth century. Although she begins with plants native to England, such as the dandelion (plate 1), the red poppy (plate 2), and the garden cucumber (plate 4), further into the project she began to expand her labors to include exotic specimens, such as nutmeg (plate 353), cinnamon (plate 354), the "Coco-nut Tree, Cacao" (plate 373), and several varieties of tea including "Green Tea" (plate 351) and "Bohea or Peco Tea" (plate 352). With the exception of the cacao, which Blackwell identifies as being used primarily to produce chocolate for gastronomic purposes, the other exotic specimens are given medical notations for their uses in important and varied apothecary's recipes. Anise (plate 374), for example, is listed as a kind of purgative, as useful in the treatment of respiratory ailments including asthma, and as an ingredient in "Linaments, particularly for the Pleurisy and other Pains in the Side."[44] In several cases, the exotic specimens were also recently imported to England and would have been new to the Chelsea Physic Garden at the time of Blackwell's incorporation of those specimens into her herbal. Occasionally, select specimens were provided to her from other locations, including the clove (plate 338), which she notes as "taken from a Branch of the Tree at Sir Hans Sloan's."[45] Evidently, prestigious scientific figures such as Sloane continued to actively support her botanical endeavors even beyond their financial support and the clout they lent to her publication.

At the same time, Sloane's support of Blackwell's botanical endeavors puts their relative valuation of plants in stark contrast. Sloane's botanical efforts were significantly motivated by the financial benefits accrued from sourcing colonial plants for botanical and medical purposes, as he took advantage of imperialist acquisition of foreign specimens to amass a substantive naturalist collection. Blackwell's participation in colonialist bioprospecting was more indirect; as a woman, she could not participate in collecting voyages, but she could interact with exotic specimens once they became transplanted into the soil of the Chelsea Physic Garden. The significance of exotic specimens to Blackwell was therefore not primarily about their financial value, but, rather, about their cultural and even emotional value. Exotic plants were important not because they could produce new pharmaceutical products or become themselves new commodities, but because they could engage scientifically interested women looking to expand their naturalist knowledge. Most important, Blackwell also esteemed each specimen for its potential role in creating a virtual community of plants and women through the publication of her ambitious herbal.

The combination of native, exotic, and acclimatized species in *A Curious Herbal* also places this botanical text in a special position among women botanizers and naturalists in the eighteenth century. Unlike Priscilla Wakefield and Maria Jacson, who preferred native specimens in their pedagogical works, or Charlotte Lennox and Eliza Haywood, who advocated for the practice of natural history in the space of a country garden, Blackwell's vision for her botanical text recognized the wealth of diverse plants being cultivated in England even before the famed voyages of James Cook and Banks. Although she limited the number of exotic specimens she could potentially include by choosing the format of the herbal for her botanical work, she was still able to include several spices, teas, and fruits, as indicated above. Additionally, unlike many early modern herbals, Blackwell's images and content are all completely unique, rendering her global approach one of her own choosing, rather than an accidental result of generations of inter-Continental copying.

Blackwell also incorporated specimens that we do not today recognize as plants; specifically, four different varietals of coral in plates 341–44. Listed as "Red Coral. Corallium Rubrum," "True White Coral. Corallium Verum Album," "White Coral. Corallium Album," and "Black Coral. Corallium Negrum," these four examples illuminate Blackwell's strategy for selecting plants for her herbal. As indicated in the accompanying text with each image, three of the coral samples were taken from "Mr. Nicholls's curious Collection," and two were taken from Sloane's "cabinet" (there were two different images for the plate illustrating true white coral).[46] As with all of her scientific details, Blackwell was scrupulous in notating her sources, being careful to note the exact specimen number of the coral samples taken from Sloane's collection. In addition to providing a layered authority for her work (Sloane's collection underpins her collection), it also provides evidence of the extensive research she undertook to produce the herbal itself. And while evolving understandings of where coral fell in universal taxonomic schemes of flora and fauna meant that coral was a plant in the eighteenth century, an insect's home in the nineteenth century, and an invertebrate polyp today, their inclusion in Blackwell's herbal represents a desire to truly provide a comprehensive resource of "curious" plants for her readers and subscribers.

Her depiction of the corals exemplifies her dedication to preserving the organic context of her vegetal subjects and celebrating that context with her readers. For example, red coral is identified as growing "upon the rocks at the bottom of the Tyrrhenian Sea and on the coasts of Spain and France," a caption that correctly identifies coral as a specimen that lives and grows under water. Yet the accompanying illustration distorts real growing circumstances by placing the corals above water and growing upon what appears to be rocks or driftwood (plate 341). However, given the likelihood that she avoided drawing the coral entirely underwater for the purposes of replication (it is easier and cheaper to use less paint in texts designed to sell many copies, especially if you are hand-coloring them, as Blackwell did), I propose that she actively sought to embrace the natural context of the coral and to celebrate its curiosity as an aquatic specimen. This strategic choice reveals her investment in teaching women about the natural world in immersive ways—in some cases, at the cost of precise accuracy. In contrast, the illustration of red coral found in plate 82 of *The Naturalist's Miscellany* (1789) avoids visual depiction of its habitat altogether and instead draws the coral attached to a seashell—either a fanciful illustration or a representation of the specimen's final, dead form as wrested from the bottom of the ocean.[47] Blackwell's illustration is thus more characteristic of the coral as a wild, undomesticated species still living, while simultaneously evoking the less technical style of the florilegia, in which backgrounds played a significant role in the composition. In this image, she blends artistic styles to create a final image that honors both the scientific requirements for an herbal and the imaginative desires of her readers.

Another individual image of note in *A Curious Herbal* is the "Fig Tree. Ficus" (plate 125; *Ficus carica,* or the common fig), a plate that reveals not only her botanical acumen and artistic direction, but also the extent of her scientific resources. Her Fig Tree illustration (fig. 1) depicts the leaves and branch of the tree, as well as the fruit in three separate stages: unripe and green, ripened, and cut open to reveal the flesh within. The color on this particular copy is especially vibrant, with vivid fuchsia and gradations of pink comprising the coloration in the fig itself. There are also specific curvatures to the interior of the fig and leaves that seem distinct to the exact plant she drew from. A comparable plate by Ehret, published in a collaborative work with Trew and Johann Jacob Haid entitled *Plantae selectae quarum imagines ad exemplaria naturalia Londini, in hortis curiosorum nutrita* (1750), reflects more symmetry in detail and more muted color on the interior of the fig. While Ehret's painting (table 73) also reflects greater attention to minute botanical details, including seeds and multiple stages of decay in the fruit itself, Blackwell's image likely reflects the circumstances under which she was able to produce the illustration. Intended as a point of reference and a faithful record of the plant as she was able to perceive it, in all likelihood her image does not reproduce microscopic views or dissected specimens because she had no access to or facilities with advanced scientific tools (cheap, portable microscopes were not yet popular). Male artists are often given greater credit for the accuracy and detail of their images, while women botanical artists could face many disadvantages in these respects.

The layout, format, and illustrative style of *A Curious Herbal* all work together to make the text as accessible as possible to a female reading audience. Blackwell's use of English and clear, concise notations made it easy for readers to understand the content and more likely that they would purchase the text. At the same time, her imaginative approach to depicting each specimen facilitated the creation of a larger scientific community in which readers of *A Curious Herbal* could envision themselves sitting alongside Blackwell in the Chelsea Physic Garden. Finally, as I

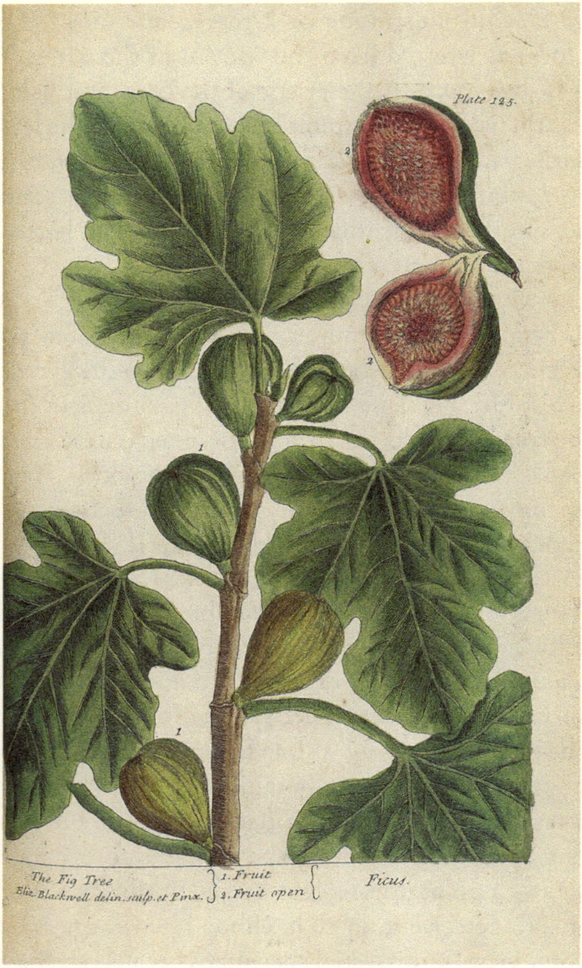

FIGURE 1. "Fig Tree," Elizabeth Blackwell, 1737. (Image from the Biodiversity Heritage Library; contributed by Missouri Botanical Garden, Peter H. Raven Library)

will elaborate below, her purposeful embrace of a botanical subjectivity in lieu of the conventional scientific objectivity represents a distinctly female approach to botanical study.

An Educator and an Artist

Little is known about the life of Henrietta Maria Moriarty. Some have speculated that she was a governess to well-to-do families of some rank; she was certainly accomplished in the usual female skills of drawing and painting, as well as botany.[48] The subscribers listed in the front of

Viridarium also suggest an unusual number of high-born connections, including the Duke of Sussex (who purchased two copies), as well as various military men, clergymen, and high-born women including Lady M[argaret] Walpole, a Lady Field, and a Lady Gardiner. Her dedication to the second edition, *Fifty Plates of Green-House Plants,* was addressed to the Dowager Lady de Clifford, whom she thanked for "kindness to me and my children."[49] While in part an explanation of her motivation, this maternal posture also indicates Moriarty's abiding interest in presenting herself as a pedagogical authority.

Moriarty's profession explains the educational nature of *Viridarium,* a text that claims at the outset to be a resource for the botanically uninitiated. *Viridarium* was also designed to assist teachers in "the very obvious use of correct drawings in public boarding schools," which reflects a twofold purpose. Moriarty's concern with both scientific and artistic education also indicates the extent to which these two subjects were very often intertwined for women in the late eighteenth and early nineteenth centuries. For her readers (and for her patrons who sponsored the work and purchased copies), it was both proper and important that young girls learn botany and drawing side by side, a convention of early nineteenth-century education that had evolved from late eighteenth-century ambivalence about botany. In fact, her text was considered appropriate for young girls in a period where so many botanical texts were not. For example, the edition of *Fifty Plates of Green-House Plants* at the Chicago Botanic Garden's Lenhardt Library is inscribed to a young Elizabeth Saxon as a gift from an older woman in November 1811, indicating that Moriarty's text was utilized as intended.

Moriarty was also the author of three novels, *Brighton in an Uproar* (1811), *Crim. Con.: A Novel, Founded on Facts* (1812), and *A Hero of Salamanca, or, The Novice Isabel a Novel* (1813). *Brighton in an Uproar* has recently been considered significant to Moriarty's own endeavors to publish a botanical manuscript because of some narrative similarities. The main character of the novel, Hubertine Mortimer, finds herself in dire financial straits after her abusive husband dies and leaves her with tremendous debt and four young children to raise. However, the always-resourceful Hubertine has a profitable idea. The narrator explains that "drawing had always been a favourite occupation with her; and she was advised to publish a botanical work by subscription. She was averse to this, as she knew her abilities were not equal to such a task; but as it was expected of her, she immediately set about it, trusting that a generous public would make allowances when they considered the motive which induced her to intrude the publication on them."[50] The narrator continues to report that, for her husband's debts, "from her exertions she paid them all within two years, amounting to the sum of four hundred and eighty pounds."[51] Whether or not her own financial circumstances echoed those of her character, it is significant that Moriarty recognized the fiscal potential within a botanical publication. While some critics have taken Hubertine's reluctance to publish because of her amateur artistic skills to reflect Moriarty's own doubts in her abilities—thereby justifying a critical dismissal of Moriarty's artistic skills—I think it more likely that such a deflection on Moriarty's part is a representation of the typical modesty topos so very common in women's scientific literature of the period. Further, given the echoes between Brunton's *Self-Control* and Moriarty's *Brighton in an Uproar,* it also seems evident that this plot device was utilized as a plausible scenario in which morally upright women might support themselves in times of distress.

A Pedagogical Project

Moriarty's project might also have been called a "curious" botanical volume for its creativity and variety in subject matter. It was published in 1806 in London under the title *Viridarium: Coloured Plates of Greenhouse Plants, with the Linnean Names, and with Concise Rules for Their Culture.* As mentioned, the work was popular enough to run to a second edition in 1807, retitled *Fifty Plates of Green-House Plants, Drawn and Coloured from Nature: with Concise Descriptions and Rules for their Culture: Intended Also for the Improvement of Young Ladies in the Art of Drawing.* Her aim, as suggested by the titles, was to provide both horticultural and artistic resources for those interested in a variety of different plants. She selected her plant specimens on the basis of their curiosity or some intriguing factor, rather than their usefulness: "I have therefore delineated a select number of such plants, as are remarkable for their beauty, their odour, or some peculiarity in their oeconomy."[52] Because her focus was specifically greenhouse specimens, her examples are all acclimatized and exotic plants, which also expands the range of her samples beyond that of many other botanical productions by women like Loudon or Pratt, who centralized plants indigenous to Great Britain in their illustrated works. *Viridarium* included plants from all over the map, including the southern part of Africa, Italy, New Zealand, Japan, China, America, and the West Indies. Like Blackwell's, Moriarty's work reflected the global reach of botany in the eighteenth century.

However, unlike Blackwell, whose *A Curious Herbal* was produced before the spread of the sexual system by Linnaeus and his acolytes, Moriarty in her project was obliged to grapple with the issue of whether to be explicit in her usage of the Linnaean system and risk offending readers or to avoid the system altogether and risk accusations of inaccuracy. While the text itself features the Linnaean binomial when labeling plant specimens, Moriarty chose to avoid any explication of the sexual system. She clarifies her rationale:

> Those who have had the instruction, or, I might say, the formation, or even the fashioning of young minds most at heart, often find it difficult to obtain representations in this most pleasing branch of natural history; on the one hand sufficiently accurate, and, on the other, entirely free from those ingenious speculations and allusions, which, however suited to the physiologist, are dangerous to the young and ignorant: for this reason I have taken as little notice as possible of the system of the immortal Linneus, and of the illustrations and comments on it; nay, I have not once named the fanciful Doctor Darwin . . . it will be pleasure and satisfaction enough for me if my own performance shall prove such an one as the rising generation can consult with safety and advantage.[53]

The reference to Erasmus Darwin will not surprise anyone familiar with the objections to his sexually suggestive verse. Moriarty's decision to "not once [name] the fanciful Doctor Darwin" is in keeping with more conservative views on women's botanical study in which such anthropomorphizing allusions to plant sexuality risked offending women or compromising their innocence. Similarly, the idea that such "ingenious speculations and allusions . . . are dangerous to the young and ignorant" represented the views of many who discouraged women from involvement in the botanical sciences. However, as explored in earlier chapters, Maria Edgeworth and Maria Jacson both approved of Darwin's work and utilized it in their pedagogical projects to teach young children of both genders about botany. Moriarty's approach is thus more cautious than that of some of her peers, perhaps because of financial instability or due to her own perspective on what might constitute a proper education for young ladies.

In keeping with prevailing ideas about the value of natural history to a woman's education, Moriarty's tactical approach to presenting botanical subjects to young women also evokes principles of natural theology.

Just as Charlotte Lennox had done decades before her, Moriarty links the study of nature to the study of religion and evaluates the worth of botanical study on the basis of how it helped young women connect with the Deity in nature. She explains in her preface: "Another object which, I flatter myself, I have in view, without presumption, is, that I may sometimes lay the foundation of a taste for the most refined pleasure that the field or garden can afford, and, at the same time, instil [*sic*] an early conviction of the wisdom in which are made the manifold works of the Supreme Being."[54] The "refined pleasures" here are both aesthetic and religious, offering multiple avenues for young women to appreciate and enjoy the natural world. By claiming to "lay the foundation" for the ability to appreciate nature as adults, Moriarty is making ambitious assertions not only about her own work, but about the value of botany to a young lady's education. An illustrated botanical reference book is therefore as integral to an early education as history or biography, all of which lead ultimately to reflections about religion and faith. In this respect, her ambitions are consonant with contemporary pedagogical theories that advocated for religious education as the underpinning of all other knowledge. Unlike Edgeworth, whose pedagogical projects imply but rarely specifically mention religion, Moriarty places her dedication to faith front and center in her text.

However much Moriarty approved of botany as a component of a rational and moral education, she was careful to modify any botanical material, because *Viridarium* was intended as an educational reference work for young ladies, rather than a work for adults alone (as was the case with Blackwell's endeavors). As she admits in her preface of the difficulty of "[obtaining] representations in this most pleasing branch of natural history," which did not contain reference to the idea of plant sexuality, her polite refusal to engage with any Linnaean theories is unsurprising. However, this decision to take "as little notice as possible of the system of the immortal Linneus" has important ramifications for the way we evaluate her work in the larger history of botanical illustration. While it is true that not all botanical artwork from the mid-eighteenth century onward directly illustrated or explained the Linnaean sexual system, it is also a fact that the most successful works at least alluded to the basic reproductive parts. Moriarty's refusal to do so has undoubtedly impacted her legacy as a scientific illustrator and author and has unfortunately meant her erasure from many histories of botany with few exceptions.

Moriarty has also been largely ignored in the history of botanical artwork on the basis of her artistic ability and perceived lack of integrity. While Blackwell sketched nearly all of her five hundred entries from live specimens in London's Chelsea Physic Garden, it is often thought that Moriarty copied her images from other sources—usually Curtis's *Botanical Magazine*. James Britten, for example, in the 1917 volume of *Journal of Botany*, accused her of plagiarism, noting reprovingly that "the plates have in almost every case been adapted with slight alteration and no sort of acknowledgement from *Curtis's Botanical Magazine*. I have not checked each one, but have compared a sufficient number to warrant this conclusion."[55] It is worth pointing out that while most botanical artwork in the eighteenth century claiming to be "drawn from nature" in fact encompassed a range of ideas about how naturalist representation was achieved (most artists learned to draw plants by copying other images and most botanical images were composites), Moriarty seems to have been especially susceptible to claims of plagiarism (not implausibly on the basis of her gender).[56] A full review of her *Viridarium* as a complete project, however, reveals that she worked hard to present a textual authority as a botanical artist that was linked specifically to negotiated ideas of originality.

For example, the very first plate in *Viridarium* is Agapanthus (*Agapanthus umbellatus*) or blue lily (fig. 2). The image itself is suggestive rather than granular, with brush strokes providing some texture and shading without the precision of attempting to copy exactly from nature. The composition is simple but compelling, with one darker green leaf arching behind the blossoms to complement the spray of light blue color at the top of the image. There is an aura of dislocation to the image as well: the stem cuts off abruptly at the bottom of the page, and the large leaf floats behind detached from the main image. This sense of fragmentation echoes the contemporary trend in botanical illustration to replicate only the plant's most significant parts, rather than accurately represent the singular plant in a given circumstance. Moriarty's Agapanthus is thus a fairly reasonable depiction of the specimen, consistent with prevailing trends in scientific illustration and, on the whole, sufficiently executed.

However, plate 500 in volume 14 of the *Botanical Magazine* (1800) seems an obvious source for Moriarty's image (fig. 3). Curtis's Agapanthus was of course done before Moriarty's, which confirms speculations that her work was copied (at least in part) from this long-running magazine. The image in the *Botanical Magazine* reflects an almost identical composition with the singular major difference being a reversal in the direction

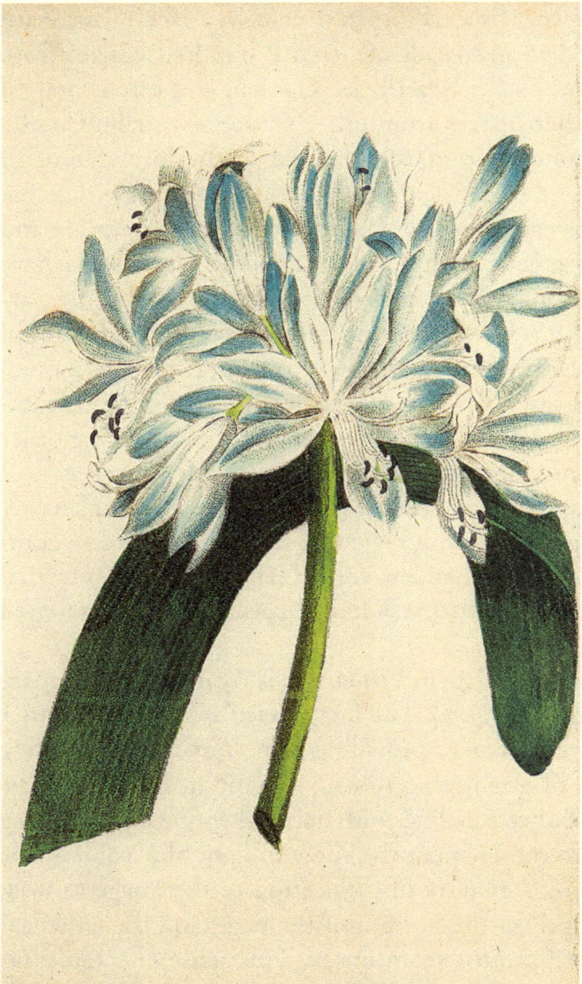

FIGURE 2. "Agapanthus," Henrietta Maria Moriarty, 1806. (From the Rare Book Collection of the Lenhardt Library of the Chicago Botanic Garden)

of the background leaf (it now stretches from right to left). The colors are similar, albeit brighter in Moriarty's text—which of course would vary between copies in the age of manual production—and the number of blossoms relative to the stem is comparable. Curtis's Agapanthus displays a finer level of detail with regards to the linework and coloring, with greater subtlety in gradation and a more precise sense of shade and hue.

Moriarty repeats this trend in plate 37, Neapolitan Mezereon (*Daphne collina*). Like the Agapanthus, the Mezereon is brightly colored and engaging, with essential details about the plant roughly drawn in. The in-

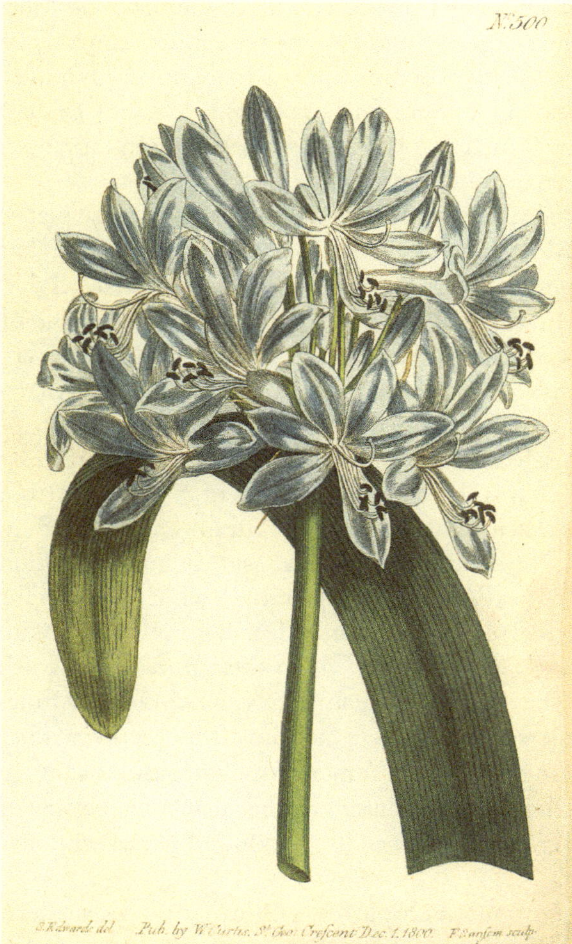

FIGURE 3. "Agapanthus," Sydenham Edwards, 1800. (Image from the Biodiversity Heritage Library; contributed by Missouri Botanical Garden, Peter H. Raven Library)

clusion of fully opened flowers as well as a blossom in the process of opening as a design choice also adheres to dominant standards of botanical illustration, in which the temporally dislocated plant image represents the multiple stages of plant growth rather than a distinct, individual plant specimen. The composition is well balanced on the page, and although the stem ends abruptly it does so in a visually appealing manner. Significantly, the *Botanical Magazine* also published the image of a Mezereon in plate 428 in 1798. When comparing Moriarty's later image to Curtis's Mezereon, the composition has been flipped on the vertical axis; the unfinished blossom is now on the left, and the curvature of the stem has been reversed. The colors are less vibrant in the *Botanical Magazine*, but the detail is more specific and the linework more refined. The flowers at the very top of Curtis's composition are more nuanced in their depiction, and the engraving is more finely executed. It seems evident here again that Moriarty's image of the Mezereon has essentially been copied from another image, rather than from life.

A review of the textual apparatus accompanying many of these plates, however, suggests that Moriarty was actively concerned with the idea that she was intended to replicate the images from live specimens. Recall that the second edition of the volume is entitled *Fifty Plates of Green-House Plants, Drawn and Coloured from Nature,* which affirms from the very first that Moriarty claims to have copied her flowers from live growth examples. She continues to make claims for originality within the descriptions of the plates themselves. For example, plate 3, the Partridge Breast Aloe (*Aloe variegata*) contains the very specific note that "the plant from which this drawing was taken, was two feet high, but in general it seldom exceeds eighteen inches."[57] She includes a similar aside in her description of plate 4, the Marsh Bastard Alum (*Calla palustris*), which reads "This drawing was taken from a plant nearly three feet high, it blossoms in March, and continues in leaf all the year."[58] Do these notes distinguish plants that she copied from life, rather than from Curtis? Do they function to reassure her readers that she did in fact copy from life, even if her compositions look strikingly familiar? Why include such asides at all, in a text that is otherwise unspecific about where her plants were sourced from (other than the fact that they were all to be found in a greenhouse of some kind)?

In answering these questions, multiple observations might be made. It is significant that Moriarty chooses plants that group together in multiple ways: they are all sourced from a greenhouse, the bulk of them blossom in the summer, and all of them are exotic specimens with some degree of success in acclimatizing to controlled environments in England. Given the likelihood that she had supporters in high positions, it is a distinct possibility that she had access to a private greenhouse of some kind. Her visits there could have coincided with the summer months (hence the reason nearly all of her examples blossom between June and August), and she could have had repeated if not consistent access to the plants themselves. Her choice to produce a text focused specifically on greenhouse plants was likely influenced by the types of specimens to which she had reliable access: this was a departure from her predecessors, who typically preferred native plants or plants that could be grown without too much special care and attention. It is also important to note that without the kind of elite patronage enjoyed by Blackwell, Moriarty was likely limited in her ability to be radically innovative, as her precarity ensured the necessity of adhering to established standards. Perhaps she believed that imitating an existing authoritative text was the best way to facilitate her conformity to those standards.

It is also true most of the images are not copied from the *Botanical Magazine*. Plate 19 of *Viridarium*, the Indian Chrysanthemum (*Chrysanthemum indica*) evokes an image from the *Botanical Magazine*, but the clear echoes of a copyist are not present (fig. 4). Curtis's image (fig. 5), plate 327 (1795–96), has more unfinished blossoms, a smaller bud, differently shaped leaves, and markedly different coloring. The similarities between these two images could easily be attributed to the fact that the artists were copying from the same plant in nature, which, even accounting for differences between specimens, would have looked similar enough to inspire illustrations that evoke one another. Her efforts to produce the illusion of authority and authenticity are then evidently genuine efforts, rather than an attempt to bamboozle her readers, subscribers, and patrons into thinking she produced an original work. Here I want to evoke Melissa Bailes's idea of "collective originality," the notion that women's scientific and creative labors could actively and frequently incorporate outside sources, influences, and texts while still remaining a hybridized original intellectual production.[59] Moriarty does not produce her scientific text sui generis, but neither does she replicate existing texts without any intellectual or artistic labor of her own.

In fact, to produce her text, she must have conducted significant scientific and horticultural research. Alongside the name of each plant in English and Latin, she includes the class and order according to the Linnaean system, directions for how to successfully grow each specimen in an English greenhouse, and information about the growing and flowering season for each plant. Some of the information could have been drawn from Curtis's *Botanical Magazine*—and she references Curtis's work directly in the description to plate 20, Cotinus-Leaved Dais (*Dais cotinifolia*)—but other information may have come from additional sources. If she did indeed have direct access to a greenhouse, she probably also consulted the gardener responsible for her subject plants for more information. And, although she disavows Erasmus Darwin's creative work, she could also have consulted the translations of Linnaeus's work funded

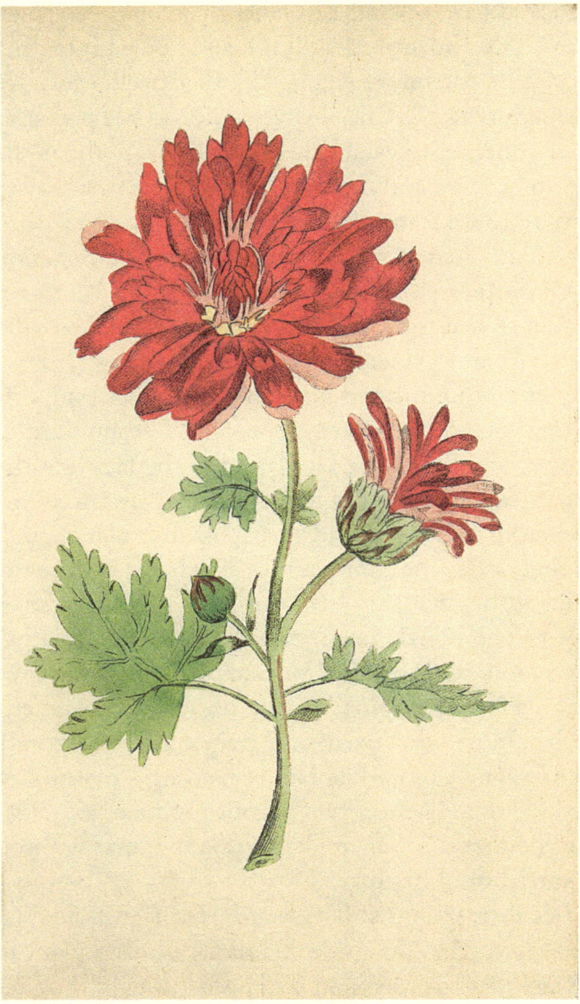

FIGURE 4. "Indian Chrysanthemum," Henrietta Maria Moriarty, 1806. (From the Rare Book Collection of the Lenhardt Library of the Chicago Botanic Garden)

by his Lichfield Society. (She did, after all, find it impossible to avoid including at least some Linnaean data in the production of her work.)

Moriarty's text is also accompanied by multiple indices in the back of the volume, including an index in which the plants are listed in English, one in which they are listed in Latin, and one in which plants are grouped according to their class. Here is greater evidence of her scientific aspirations: these indices link her work to other horticultural and botanical reference works and further differentiate it from illustrated manuals and drawing aids for young ladies. Indeed, in spite of her delicate avoidance

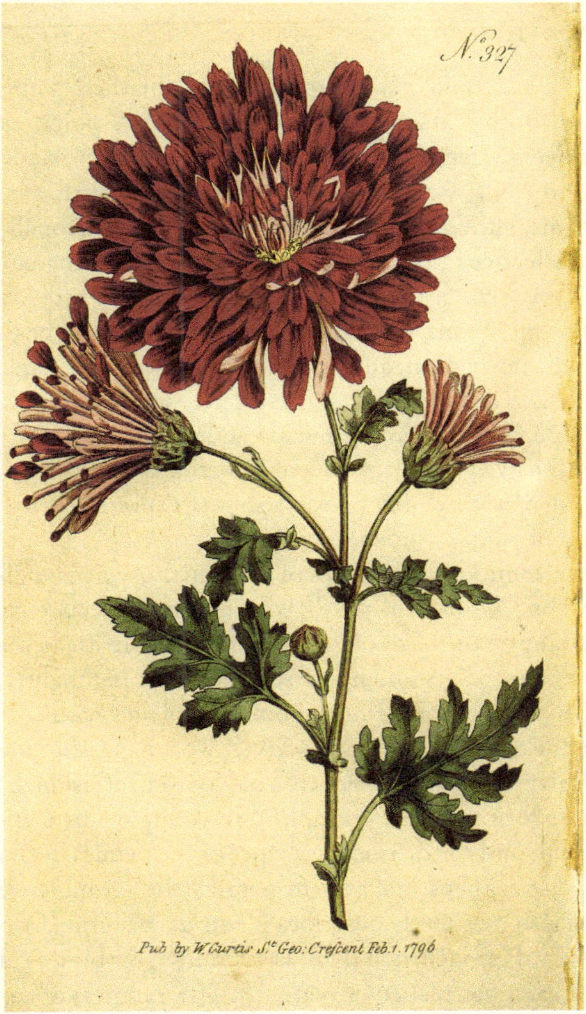

FIGURE 5. "Indian Chrysanthemum," 1796. (Image from the Biodiversity Heritage Library; contributed by Missouri Botanical Garden, Peter H. Raven Library)

of certain elements of the Linnaean system, Moriarty utilizes the classification system faithfully and accurately. Taken together, the different botanical frameworks—order/class divisions, observations of growth patterns, horticultural directions, and scientific indices—of *Viridarium* suggest substantial intellectual labor on Moriarty's part. Her reference to and occasional copying from Curtis's *Botanical Magazine* must therefore be seen as part of a larger pattern of research consumption. She adapts both printed and visual information from the *Botanical Magazine* in the production of her own, original botanical work.

Horticultural Intimacies

A Curious Herbal and *Viridarium* were also significant for their ambitious pedagogical goals. Both texts were designed to serve as points of reference—with intended audiences ranging from professionals to amateurs, from adult men to young women—and both were ultimately successful in their intentions, although Blackwell's *Herbal* has traditionally been considered more effective in the long term. Like the many cognate publications produced by male artists at the time, these two projects successfully entered the marketplace of educational books on the basis of their clarity, depth of information, and accuracy of illustration. Aside from their status as female-authored texts, these reference works were also notable for their specific targeting of female audiences: Blackwell's herbal surely included the lively tradition of female herbalists, midwives, healers, and cooks, and Moriarty's later textbook was expressly designated for the education of young women.

There were, however, some key differences in the way those pedagogical aims were accomplished. Unlike Blackwell, whose plant subjects were all selected at least partially on the basis of their practical uses in medicine or cooking, Moriarty embraced a standard of selection that highlighted something she found personally compelling about each plant's nature—visually, emotionally, or viscerally. Whereas Blackwell, working within the herbal genre, selected plants for their purported utility to professionals, Moriarty selected plants for their aesthetic appeal to young women. Indeed, the vibrancy of color and the distinctive shapes of the petals chosen in each of her included specimens make them noteworthy on multiple aesthetic registers. She also frequently describes plants as "beautiful," reflecting a different standard for inclusion than was common for botanical reference texts (see, for example, plate 6 or plate 10). The "richness of colour" is noted multiple times as well, which is evidenced in the vivid coloration on her plates: plate 23, Rigid-Leaved Gorteria (*Gorteria rigens*) is a particularly fine example of scarlet and orange pigmentation. We might also think of the greenhouse, as a nonprofessional zone, as a garden space more accessible to female botanists: the character of Hortensia in Jacson's *Botanical Dialogues,* for example, cultivates a personal greenhouse. By situating her botanical text within a genre of garden that had fewer restrictions, both financially and logistically, and less of a singular association with male botanists, Moriarty creates a representation of botanical knowledge with the potential to function outside of professional spaces like the Royal Botanic Gardens at Kew or the Chelsea Physic Garden.

Promoting broader and more accessible garden spaces in her text, she thereby also specifically enables women to interact with and learn from her botanical reference text.

Likewise, Blackwell was actively interested in promoting her text for female readers. Although *A Curious Herbal* was designed for a professional audience, the information contained within is also basic enough to appeal to those without academic medical training—implicitly including the large populations of women who were trained in medical traditions outside of the newly codified professional class. She also makes reference to specific medical treatments that evoke culinary practices, like "oils" or "decoctions," recipes that could likely be replicated at home with some additional knowledge, such as the relative proportions of ingredients and other inclusive components. Therefore, while *A Curious Herbal* could never be a sole resource for anyone seeking to learn about an herbal remedy with no prior knowledge, male or female, it provided a useful point of reference in conjunction with other, more detailed reference texts or robust experiential knowledge. Given her avoidance of Latin, her abstention from too much technical terminology, and her decision not to include detailed medical notes or recipes, Blackwell's text was accessible to a broad audience.

However, *A Curious Herbal* and *Viridarium* were important for more than their accessibility to female naturalists. Blackwell's and Moriarty's purposeful cultivation of a direct relationship with unique gardens and specific plants also suggests something significant about women's botanical artistry that is rarely discussed in the history of women's participation in the sciences. Although, as Judith W. Page and Elise L. Smith have demonstrated, gardens in the eighteenth century were often conceptualized as a space for pedagogical development and female learning, the value of gardens to a vibrant history of women naturalists is often underrated.[60] As I have suggested above, botanical gardens had aesthetic dimensions just as pleasure gardens possessed scientific dimensions; both might be understood as overlapping spaces of intellectual inquiry and emotional satisfaction. But how does such a construction inform scientific practice? Given that we understand gardens to be spaces that summon imaginative conceptions of "invitation and enclosure," as Ross puts it, botanical labor in a garden is at once an expansive and a limited practice.[61] Women undertaking scientific endeavors in the space of a working professional garden like the Chelsea Physic Garden or a pleasure garden like a private greenhouse must then grapple with all of the metaphorical associations between women and gardens as they become literally and figuratively invited into the garden space and then enclosed within it.

Women like Mary Wollstonecraft, Maria Jacson, and Priscilla Wakefield were able to productively rewrite metaphorical associations between women and plants in their intellectual productions; Blackwell and Moriarty accomplished an equally impressive reversal of similar metaphorical links between women and gardens. Where the idea of women as delicate, beautiful flowers was difficult to combat, the affiliations between women and gardens were perhaps more challenging. Ross explains, "Like a woman, nature was a source of pleasure; like a woman, she was to be used and improved; and, like a woman, she was to be available to her master yet chastely screened from the view of others."[62] In producing botanical reference texts that adhered to dominant standards of scientific illustration while simultaneously celebrating a personal connection to the organic life within their subject gardens, Blackwell and Moriarty invert the proposition. Nature was a source of pleasure and instruction to women, but the pleasure was a different kind altogether—not one of dominion, in which nature was a space to be "used and improved," but one of respect, wherein nature is an ally and a companion, an artistic subject and not an object. By producing botanical works that remained not within the confines of a home or in circulation among intimate female friends but were instead published and widely disseminated among a mixed audience, these women also resisted the idea that nature was to be "screened from the view of others."

These publications also destabilize the idea of gardens as fundamentally enclosed spaces. While both Chelsea Physic Garden and Moriarty's private greenhouse were literally and figuratively enclosed land, by producing artistic works that lovingly captured and then carefully distributed the images of the plants within, both women place greater value on the invitational aspect of gardens. Inviting their readers visually and metaphorically to enter these gardens with them, Blackwell and Moriarty create a shared space in which those readers could learn about their subject plants while also enjoying the imaginative relationship with those specific gardens. Where most professional botanists sought to produce authoritative information about plants, and contract artists like those who illustrated the *Botanical Magazine* created pieces designed as a universal point of reference, these two amateur artists produced work that must be primarily understood as disinterested in serving as an ultimate botanical authority. As reference works, they served a basic consulting purpose; as original artistic and scientific productions, they represented a distinct perspective on the relationship between women and vegetal life.

This distinctive perspective is achieved through the cultivation of a female botanical subjectivity. By embracing a subjective position in relation to their botanical subjects, both Moriarty and Blackwell open up new possibilities for the ways in which their women readers might relate to nature. A subjective stance with respect to plants is also a respectful one: a dynamic founded upon admiration and curiosity rather than exploitation and dominion, such subjectivity ultimately leads to the growth of new relationships between humanity and vegetality. Their pedagogical intentions are thus among the most radical of the period—to these author-illustrators, knowledge of plants was about more than use value. Echoing the calls to action of many botanists and popularizing authors, these women promoted the idea that the best way to learn about nature was to immerse yourself within it. Simultaneously, however, they argued that such immersion could be accomplished through an imaginative association with the gardens captured in *A Curious Herbal* or in *Viridarium*. By perusing these texts and embracing for themselves the critical value of botanical subjectivity, Blackwell's and Moriarty's readers can cultivate personal relationships with gardens throughout England, ranging from their own kitchen gardens to the Chelsea Physic Garden.

(Un)domesticated Nature(s)

I want to return to the idea of botanical gardens as both epistemic and aesthetic spaces, wherein scientific knowledge and artistic principles coexist. When we look at the history of women like Blackwell and Moriarty, whose intellectual labors were significantly shaped by such gardens—directly or otherwise—it is important to think about the ways in which the evolving status of gardens in the eighteenth century influenced their scientific labors. While most large-scale gardens in the eighteenth century, and especially those that endure in some form today, were designed and executed by men, it is vital to remember that women's sociocultural relationship to plant life and to garden spaces informed their scientific work in different ways. Significantly, while "women writers and artists used the subject matter of gardens and plants to educate their audience, to enter into political and cultural debates . . . and to signal moments of intellectual and spiritual insight," the tension between such critical deployments of the garden space and the knowledge structures that underpinned that same concept beg further exploration.[63] Likewise, where some criticism has interpreted women's usage of the garden idea in fiction or in art as a replication of certain types of domesticity, it is vital to understand the extent to which some women actually produced radical models of the relationship between themselves and plants that disrupted, rather than sustained, domestic ideals.

While well-respected professional artists like Ehret and Catesby produced paintings with nearly lifelike realism and a usually strict adherence to scientific standards of illustration, Blackwell and Moriarty, as amateur artists, cultivated a relationship to their subjects that reflected their de-emphasis on exactitude in favor of reinterpretation. Blackwell was scrupulous in documenting her sources, even to the specimen number (when not taken from the Chelsea Physic Garden), but her final productions suggest a faithful dedication to particular specimens rather than adherence to botanical norms. Likewise, Moriarty's own botanical labors were more of a collective effort, an homage to existing artistic sources alongside her own individual productions. In each case, the artist also subtly emphasizes the extent to which the highly structured relationship between painter (actor) and plant (subject to be acted upon) possesses a complex dynamic. For Moriarty and Blackwell, plants are not endlessly replaceable models for a painting that might serve as universally representative: instead, they are individual pieces of organic matter with their own unique characteristics and qualities.

Part of this emphasis on plant individuality comes from their sourcing. Both Blackwell and Moriarty produced their illustrated reference books using a particular garden as a source for their reference specimens: the former, the Chelsea Physic Garden, and the latter, an unidentified greenhouse, possibly belonging to a patron or employer. With consistent access to a professional or expansive private garden, each of these women enjoyed a kind of elite position that amateur female artists rarely occupied. The need for this kind of privileged relationship to exist in order to produce the kind of high-level scientific work that Blackwell and Moriarty put forth could potentially explain why more women did not undertake major botanical projects in the eighteenth century. This personal relationship to an individual garden also signals something very important about the ways in which women as naturalists inhabited a unique position—that is, by cultivating, in more ways than one, an intimate relationship to a specific garden, these women propose a productive allegiance with vegetal life.

Rejecting contemporary standards for botanical illustration, which essentially demanded an unrealistic representation of a plant for the sake of scientific thoroughness, both Blackwell and Moriarty produced often individualized images that were adjacent to but also different from those standards. Allowing the plants within their specific garden to inspire their artistry and guide the composition of their reference texts, they produced creative works alongside organic life. For example, with periodic reference to the individual plant itself, such as a mention of how tall the specimen was or how proliferating the blossoms, Moriarty refuses to elide the real plants behind her illustrations. Such specific details not only allow us to envision a real plant situated within nature (in contrast to but also in conjunction with the published image of the severed plant), but also to imagine ourselves in that garden alongside Moriarty, noting with pleasure that the plant almost reaches our waist, or that the leaves are particularly fine and plentiful.

Likewise, by occasionally including insects on her images, such as Cross-wort (plate 76) and Meliot (plate 80), Blackwell evokes a specific plant in a specific moment in time rather than offers a composite image that represents a theoretical ideal. While insects were often used as a point of reference in botanical images (to provide an idea of scale, or to give a sense of the plant's native context), here the inclusion of creatures like the caterpillar must also evoke metaphorical associations with femininity. Small, delicate, and often considered beautiful, these minute creatures signify in part the same kinds of limitations—of movement, of emotion, and of mind—that women in the eighteenth century almost always faced. Just as Charlotte Lennox utilizes images of insects to not only teach her readers about natural history but also to gesture toward the potential freedoms that such intellectual labor could hold, Blackwell's inclusion of tiny insects in her botanical plates implies the greater possibilities inherent in the study of botany.

I want to emphasize, moreover, that such a nod to domesticity does not imply that the garden serves merely to reinforce domestic ideals upon women as plant-like subjects, passive and sessile. Rather, this inclusion of small insects alongside large blossoms highlights connections between people and the natural world, reinforcing the notion from contemporary feminist environmental studies that nature can function as a critically mutable space with particular potential for women. As Stacy Alaimo explains, "Women, of course, do not stand outside the conceptions of nature they seek to transform. Nor do women enjoy some sort of special relationship to nature . . . [but] women are positioned differently within gendered discourses of nature that usually work to their disadvantage" and must consequently work to "reconfigure these discourses."[64] For eighteenth-century women like Blackwell and Moriarty, the reconfiguration of such discourses depended upon the recalibration of perspective.

This shift in perspective involves the transition from utility to curiosity. Neither Blackwell nor Moriarty was in a position to exert control over the composition, growing conditions, or aesthetic dimensions of their respective gardens. Instead, they devoted themselves to capturing the essence of the individual plants found within each garden, each of which are "curious" in their own right. While Moriarty did not allude to the concept of curiosity in her title, she selected her greenhouse specimens for something distinctive about their color, scent, or growing circumstances. Blackwell's deployment of curiosity weaves throughout the volume, as we see exotic specimens interspersed with common indigenous plants, each framed as worthy of inquiry and interest. By assuming a position of admiration and respect for the individual qualities of their vegetal subjects, these two artists change the typical relationship between painter and painted. And, although they did not diverge significantly from canonical standards of botanical illustration, rendering their work at least visually similar to that of male artists, their ability to work within those conventions while celebrating the individuality of plant life remains impressive. Further, as Bending has argued, when we recognize gardens as providing "the opportunity for a self-fashioning engagement with cultural norms and narratives," we reveal entirely new possibilities for critical interpretation of women laboring within botanical gardens.[65] While Bending is specifically writing of women as gardeners with creative and ecological control over plant specimens, I think it evident that women as artists working within those gardens were equally faced with "the disparate agenda of eighteenth-century culture."[66] Blackwell and Moriarty thus produced in their botanical representations of vegetality not only a faithful record of plant specimens, but an invaluable representation of how they negotiated such gendered spaces.

The early runs of the *Botanical Magazine* provide a useful point of contrast with these female-created botanical illustrations. At this point in the publication's history in the late eighteenth century (while still under the editorial control of William Curtis himself), the illustrations were all done by men, many if not all of them by Sydenham Edwards and James Sowerby. These images are accurate, well-colored representations of the specimens, and they provided a standard for botanical illustration that women like Moriarty would work to follow carefully. However, most of the plants in the *Botanical Magazine* during the eighteenth century were plants of European extraction: there were few samples of plants from, for example, the Americas, the West Indies, and Asia, in spite of the British/European colonial or mercantile presence in these areas. While this would shift in the nineteenth century as public interest in exotic botany grew, during the time of Moriarty's publication she was including a greater number of exotics than the *Botanical Magazine*, reflecting both a narrower and more forward-thinking botanical focus on her part. Further, the illustrations in these early runs were not painted by Edwards or Sowerby; rather, they were completed by hired colorists who were given little pay and no credit (and who may have very likely been women). Blackwell, conversely, is known to have hand-colored her own engraved images, participating in multiple stages of production where male artists typically provided the initial image alone. We do not know enough about Moriarty's work to know if she colored the images herself; the degree of imprecision in the coloring is consistent with the quality of the drawing and may suggest her participation at both stages. Finally, the images in the *Botanical Magazine* in this period are never sourced; readers are not told where the specific plant specimen the image is taken from is to be found. With the exact attribution of the Chelsea Physic Garden in *A Curious Herbal* and the implied specificity of a particular greenhouse in *Viridarium*, the work of these women ties directly to explicit spaces and distinctive plants.

There are some essential differences between gardens and greenhouses, botanical spaces that are themselves distinct from what is typically thought of as that "undomesticated" nature that Alaimo writes about. Whereas gardens as least ostensibly imagine themselves as organic spaces in which nature is allowed to flourish as it will, within certain aesthetic and physical boundaries, greenhouses are fundamentally places of dislocation, separation, and production. Even in the eighteenth century, greenhouse plants were typically grown in discrete sections—this varietal here, this species there—that were organized by connections at once highly functional and delightfully arbitrary: plants requiring a heating component located in this corner, all the purple flowers located in that corner. Today, plants in a greenhouse are often grown for sale and therefore distribution and disruption, whereas gardens are usually intended to remain whole. Greenhouses make exceedingly visible what Deidre Lynch calls the "strange intimacy" between exotic and domestic flora, whereas gardens often attempt to soften and naturalize that intimacy by themed beds or areas.[67]

All of this is true. Yet I want to point out that there are greater underlying connections between these two botanical spaces than we might at first assume. Both spaces collapse the physical and metaphorical distance between "point of origin" and "current growth location." Even indigenous plants like dandelions or clover in a place like the Chelsea Physic Garden have been taken from their natural habitat and purposefully planted within this space. Exotic specimens have traveled even farther, of course, but they, too, have been displaced. Both greenhouses and botanical gardens require a certain amount of finesse in keeping select species alive; there is often just as much manipulation of soil, temperature, and other growing conditions within gardens as there is within greenhouses. Even more important, both spaces are reliant upon "artifice and exhibition" to perform their intended function: the "natural" aspects of both gardens and greenhouses are truly, at the most basic level, unnatural.[68] So, where does this leave women botanical artists who seek these spaces out as their points of access into natural and scientific discourse?

Ultimately, the constructed-ness of both botanical gardens and greenhouses problematizes any argument that might be made about the relationship between women and plants. Yet, simultaneously, the fact that we can recognize either space as an artificial one also destabilizes the fiction of "nature" as a nonartificial space. Nature is always inflected with the associations and assumptions brought to bear by human beings; even in the wildest of places, nature is not a neutral site. As contemporary environmental feminists have argued (from varying perspectives and with sometimes wildly different conclusions), "nature" is at once a constructed, cultural space and an organic, ecological space.[69] Likewise, Alan Bewell in *Natures in Translation* (2017) points to the need for critics to understand "nature" as both a pluralistic concept—there are always multiple "natures"—and as a culturally contingent and mutually constituted biological space.[70] For Blackwell and Moriarty, carefully cultivated professional or private gardens were invaluable forms of nature/natures that facilitated their connections with and contributions to the global circulation of organic specimens, botanical images, and printed scientific texts.

At the same time, their access to organic plant matter was deeply intimate, corporeal, and immediate. Imagine Blackwell sitting among the plants and insects, perhaps in the dirt herself. In order to paint the plant with the type of detail displayed in plate 374, she must have gotten quite close. She must have touched the plant, and she must have done so more than once. She must have smelled the distinctive odor of the raw anise, the other plants around her, and perhaps her own body. She and the anise plant breathed the same air for a time, occupied the same space for a time. By adopting this sympathetic—and in some ways, empathetic—stance toward plants, Blackwell rewrites the typical narratives about the relationship between human and vegetal life. By utilizing this female botanical subjectivity to present her own kind of scientific authority, Blackwell encourages her readers to cultivate similar, personal relationships with plants.

Finally, while prevailing paradigms concerning the relationship between women and nature have traditionally been seen as reifying stereotypical ideas about differences between the sexes, a critical interrogation of Blackwell and Moriarty's interactions with gardens presents a different perspective. By framing their work as at once morally improving, scientifically valuable, and a source of personal pleasure, both women undermined their prefatory caveats. Their texts were not the productions of untutored women whose work would falter without the active and substantive assistance of male botanists. Rather, *A Curious Herbal* and *Viridarium* were vibrant, original works that celebrated plants—and women. If we are to learn to read women's botanical labors in the eighteenth century as significant not merely for their derivations from conventional standards but for their reimagining of those standards within acceptable boundaries, contextualizing these illustrated reference works within the framework of the garden is an important first step.

Both Elizabeth Blackwell and Henrietta Maria Moriarty produced illustrated botanical reference works that are worthy of greater attention than they have been given: on the basis of their artistic abilities, scientific acumen, and relationship with their vegetal subjects. By cultivating personal, subjective relationships with their working gardens and the plant specimens within, each of these women produced original work that occasionally diverged from contemporary ideas about what botanical illustration could encompass and accomplish. Moreover, by making these distinctive illustrated texts accessible to a broad audience that included women and young people, Blackwell and Moriarty underscored the pedagogical significance of botanical study. Finally, as naturalist-artists, these two women were also part of an important legacy of female scientific labor that spanned the entire eighteenth century—a legacy that included the remarkable botanist, artist, and crafter Mary Delany, the subject of the next chapter.

5

Collecting and Creating

Mary Delany's Naturalist Career

> [Sir John] has given me some very pretty shells for my works, but none fine enough for my collection; but a friend of D.D. in the West Indies has promised me great curiosities.
>
> —Mrs. Delany to Mrs. Dewes, February 25, ca. 1743/44

> I wish my tedious description has not tired you, but I was so pleased with the flower, &c., I could not help communicating it.
>
> —Mrs. Delany to Bernard Granville, December 17, 1771

While painting in a variety of media was the most common form of depicting plant life in the eighteenth century, botanical artistry encompassed a wide range of activities and projects for women, many of which were linked to domestically endorsed forms of handiwork or feminine accomplishment. The exceptional artistic career of Mary Delany illuminates the impressive breadth and profound creativity possible in a woman's engagement with plant life. In this chapter, I offer a survey of Delany's career as a naturalist artist—a career that complements the more conventional botanical artwork studied in the previous chapter while also participating in the same conversation about vegetality and femininity woven throughout this book. I argue that her body of work reflects not only a remarkable ingenuity with respect to generating new forms of botanical art, but also a deeply sympathetic connection to the natural world that combines the piety of natural theology with a pure passion toward plants, shells, and animals, thereby revealing further possibilities for woman-plant-insect entanglements.

In October 1772, Delany wrote to her niece Mary Port about an extraordinary new craft project she had just devised: a botanical collage made of paper, paint, and leaves. Although in this letter she writes only that "I have invented a new way of imitating flowers," she would later refer to these innovative art pieces as paper "mosaicks."[1] These mosaics were an almost instant success, eliciting avid interest and high praise from friends and acquaintances, professional botanists, naturalist explorers, and even royalty. While this collection of nearly one thousand mosaics has been housed since the nineteenth century at the British Museum—where a scant few samples are on display—Delany's work has only in the last several years garnered attention for the unusual process by which she created those mosaics and the elite botanical circles in which she moved.[2] A few of these studies, like those collected in *Mrs. Delany and Her Circle*, probe the relationship between her artistic pursuits and scientific interests through a careful accounting of her various projects. I build upon these productive studies to further illuminate how profoundly her artistry was integrated with vegetality not merely as a point of personal or academic interest, but as a deeply personal commitment to understanding nature. This chapter proposes that the interconnected nature of Delany's material craftwork and her active pursuit of natural history, as expressed through her correspondence and poetry, highlights her distinctive relationship with the natural world and the unique role that plants themselves played in her artistic productions.

Furthermore, while other critics have repeatedly toyed with the question of Delany as an "artist" or as merely "artistic," I evaluate her productions not for their place in a creative canon, but, rather, for their material function as scientific objects.[3] By looking at both Delany's renowned mosaics and the broader array of botanical and organic craftwork she created as part of a Bluestocking community, this chapter argues that such material labor represents a distinctive approach to scientific study. As Delany and others undertook projects of craftwork that incorporated everything from shells and plants to fossils and minerals, they transformed natural specimens into valuable objects of decoration and exchange in ways unanticipated by typical botanizing manuals. For Delany and her companions, such craftwork created points of access into a botanical discourse: as Ariane Fennetaux has argued, "Crafts could be appropriated by women, often allowing them to trespass into fields not traditionally associated with them."[4] Where most botanical publications at the end of the eighteenth century encouraged the study of botany on strictly Linnaean terms, these women undertook creative projects that offered distinct possibilities for intellectual fulfillment in the spaces where science and art converged.

The substantial body of nature-inspired craftwork from the eighteenth century is a fruitful archive for the study of female scientific endeavors in the period. As texts produced by, exchanged by, and collected and preserved by women, the scientific crafts of the Bluestocking circle represent a crucial model of exchange that functioned both within and beyond the elite culture of natural history to which these women were exposed. An active correspondent with Jonathan Swift and friend to Joseph Banks, then-president of the Royal Society, natural philosopher Daniel Solander, botanical artist extraordinaire Georg Ehret, and Philip Miller, the longtime proprietor of the Chelsea Physic Garden, Delany moved in rarefied scientific circles alongside her lifelong friend, Margaret Cavendish Bentinck, Duchess of Portland. While the Duchess of Portland is today regarded as the more scientifically inclined of these women, Delany's investment in the subject matter was perhaps equal to that of her companion, albeit differently expressed.[5] In fact, Delany often positioned herself among the "virtû" like Solander, Ehret, and the Duchess herself, taking equal part in their collecting, sorting, processing, and cataloging of shells, plants, and minerals. Along with the Duchess of Portland and other fellow Bluestockings, Delany also executed creative projects inspired by nature in multiple artistic genres, including needlework, painting, and outdoor garden installations. Incorporating both traditional artistic materials (paint, paper, oil, thread, cloth) and found components from the natural world (plants, shells, feathers, fossils, stones), Delany's artistic undertakings meld various types of organic objects into an amalgamated final piece with multiple cultural valences. At once artistic and scientific, intimate and strange, domestic and foreign, these artistic projects embodied the complex ways in which women participated in naturalist study in the eighteenth century.

This chapter will range through Delany's natural craftwork, including the mosaics, to her poetry and other writings to highlight her remarkable approach to the pursuit of natural history throughout her lifetime. By looking at a broad survey of her artistic and creative productions, I underscore how significant the study of the natural world was to her individually, and how important scientific subjects could be to a woman's life in the eighteenth century. Given the importance of her connections and friendships with other naturalizing women in the eighteenth century, I also want to propose Delany's artistic career as an exemplar of the potential inherent in female scientific networks at the end of the century.

Natural Craftwork as Scientific Labor

As I have discussed in previous chapters, women were actively involved in the pursuit of naturalist study from mid-century onward, an involvement that encompassed everything from the casual perusal of relevant publications to dedicated study at home and the personal collection of specimens. However, the relationship between women and botanical study in the eighteenth century was complicated. While women avidly participated in the collection and utilization of plant specimens, they could not be the arbiters of natural knowledge according to most academic standards; as Londa Schiebinger notes, women "were not taxonomists, nor among those shaping the future course of the science," despite their frequent participation in botanizing activities.[6] Culturally and metaphorically speaking, women were also linked very closely with nature (especially the cultivated garden space) in sexualized and gendered ways—associating women and plants as subjects of botanical inquiry, rather than as agents of it.[7] As Asia Haut has traced in great detail in her assessment of the impact of Linnaean ideology on literary and artistic works, a woman's relationship to botanical systems in the eighteenth century was thus an inevitably fraught one in which "the feminine is at once the botanical subject and the botanical student," and, consequently, "the female may never have unmediated domain over the botanical."[8]

Of course, women naturalists actively collected, cataloged, and classified specimens throughout the eighteenth century. Just like the untold number of male amateurs interested in natural history, women participated in the material practice of specimen collecting in significant numbers and with great success. Generalized histories of science have tended to assume that structural and cultural inequalities between women and men disadvantaged women in the pursuit of scientific study, but such an analysis elides the empirical work being undertaken in the spaces of overlap between science and craftwork. In fact, the amateur and female collection of specimens botanical, conchological, and animal directly contributed to the expansion of those fields of study: as Pamela H. Smith explains, "Enthusiasm for observing and representing nature legitimated the study of nature more generally, and facilitated social acceptance of a new empiricist investigation of nature."[9] While Smith's analysis refers to early modern sites of craft and empirical inquiry, such an interpretation seems only more germane to the late eighteenth century, a period in which the significance of empiricism to scientific inquiry was continually affirmed by professional publications and pursuits.

In many ways, women's participation in collecting practices didn't look all that different from the habits of male collectors. Consonant with many literary representations of male naturalists, Eliza Haywood's "fair philosophers" engage in the physical work of slogging through mud with specimen baskets, microscopes, and other specialized equipment. Likewise, Mary Delany and the Duchess of Portland set forth on local expeditions to seek out, for instance, shells, fossils, and plants. Moreover, as Beth Fowkes Tobin has detailed regarding the Duchess's shell collecting, the acquisition of specimens relied upon a vast network of resources that included gifts from friends' estates, assistance from local fishermen or farmers, collecting in person, and purchasing from wholesalers or specialized traders. In addition to her own explorations at the Duchess of Portland's estate Bulstrode, Delville (her second husband's home in Ireland), and various family estates, Delany received plant and mineral specimens from relatives, neighbors, the Hammersmith Nursery in London, the Chelsea Physic Garden, and, at the direction of Queen Charlotte, even the Royal Botanic Gardens at Kew. Without such connections, the bulk of her artistic work could not have been completed, suggesting a more elite access to materials that most did not enjoy. However, as Charlotte Gere and Marina Vaizey have pointed out, many different types of botanical and conchological collecting crucially "provided a network of female friendships . . . which enlarged the horizons of women far beyond the concerns of fashion, domesticity, and philanthropy to which they were customarily confined."[10] Even women without the kinds of networks or resources that so benefited Mary Delany could reap the rewards of collecting.

The conditions under which women conducted botanical inquiry and specimen collecting also possessed important social dimensions. While botanical inquiry or scientific craftwork might be conducted in private, it was rarely if ever pursued alone. Tobin explains that botanizing and the craftwork that often resulted from the practice were inextricable from a certain kind of genteel sociability: "The study of natural history brought people together to observe, draw, collect, classify, and discuss the natural world. Gathering botanical specimens to make floras and herbaria, collecting butterflies, spiders, and shells, making decorative objects from natural materials such as feathers and shells, and sketching and painting images from nature were social activities that men and women of middling and upper classes did together."[11] While conservative commentators did attempt to "[clamp] down on mixed flower-gathering expeditions," women in Delany's social position rarely faced accusations of impropriety on their botanizing ventures and were free to engage in collecting parties.[12] Whether the intention was to study, to dissect, or merely to enjoy the collected samples, the act of specimen collecting allied amateurs like Delany and the Duchess of Portland with professionals like Solander and Banks in a joint experience.

Where these women collectors differed was in the subsequent use of those specimens. In keeping with prevailing scientific practices, the Duchess diverted many of her unique specimens into her private collections, later known as the Portland Museum. However, avid naturalist collecting consistently resulted in many examples of a given specimen, at which point the less desirable specimen would often be passed to friends, sold to others, or even discarded as extraneous from more elite collections. For the Duchess and Delany, however, like many other women, those shells, plants, feathers, and fossils that did not make it into the cabinets were easily converted into raw materials for domestic craftwork. While these projects were notable for their beauty and ingenuity, these objects also reveal how natural craftwork overlapped with the investigation of nature, and the way craft could therefore itself be a tool of natural inquiry—not just a product of it. By seeking out rarer or exotic specimens within broader networks, hunting them down in person, and cleaning them themselves, Delany and the Duchess were enmeshed in the very networks of production surrounding craftwork in the eighteenth century. Furthermore, many of these processes relied upon empirical abilities to decide which specimens were appropriate or worth collecting. Observational selectivity and judgment were key components to the successful conversion of collection into craft: by seeking out shells for a specific variety, shape, and color, as the two women did for a grotto they completed at Bulstrode, Delany made not only artistic decisions but scientific ones as well.

The collection-to-craft process was also enriched by the fact that many of the specimens were exotic. Plants, shells, and feathers sourced not only from other regions of England or Europe but from territories around the globe were similarly incorporated into craft projects. Delany, for example, writes often in her letters about the acquisition or viewing of new plant specimens for her various botanical projects, many from friends like Solander or Banks. In December 1771, she writes, "We were yesterday together at Mr. Banks's to see some of the fruits of his travels, and were delighted with paintings of the Otaheitie plants, quite different from anything the Duchess *ever* saw, so they must be very new to me!"[13] During this visit they were to view multiple plant depictions collected on his voyages, including an image of a plant whom she does not name, but describes in detail: "There is one in particular (the name I cannot recollect), that bears vast flowers, larger and somewhat of the appearance of the largest poppy when full blown, the leaves all fungid; the petals that are like threads, are at the calyx *white*, by degrees shaded with pale *purple*, ending with *crimson*."[14] This excitement about new, exotic plant specimens, exhibited so clearly even secondhand through her letters, also reflects a keen eye for artistic detail: the typographic emphasis on the colors (transferred from the handwritten correspondence), rather than the technical Latin terminology, suggests that her enthusiasm was in large part a result of the plant's potential incorporation into her many craftwork projects.

This attention to the artistic potential of a plant specimen was translated throughout many of Delany's works, as she executed a wide variety of artistic productions, including "oil painting, drawing, plasterwork, shellwork, featherwork, japanning, needlework, paper-cuts, and paper collages."[15] Several of these crafted objects, like the mosaics, woven screens, decorated mirrors, or embroidered furniture, remained at Bulstrode or Delville, while others were gifted to friends throughout England. One such project, which Delany refers to in passing as a "screen," biographer Lady Augusta Llanover describes in glowing detail: "This wonderful piece of work . . . is a basket filled with flowers from nature, worked in chenille, and is indescribably beautiful, the drawing as well as shading being perfect. A peacock butterfly is also equally well represented."[16] Representative of the kind of lovingly detailed work that Delany would produce throughout her life, this embroidered screen showcases her attention to tiny facets of her subject matter as well as her accomplished skills as an embroiderer. In addition to being "indescribably beautiful," the screen is also interpreted as a faithful replication of natural objects like blossoms and a butterfly. The screen, Llanover notes, had been bequeathed to a female relative of hers sometime in the years since its production, making this particular piece of natural craftwork one among many examples of the ways in which Delany used her scientific talents to cultivate female friendships.

In fact, these objects traveled primarily, though not exclusively, in female networks, reflecting an important characteristic of their production: the imbrication of domesticity and femininity with foreignness and masculinity. As Madeleine Pelling has compellingly argued, "Craft produced by elite women within the context of the sites and spaces of the duchess's museum [positioned] such works as material and narrative extensions of the broader assemblage and [mediated] the collision of the global and domestic at Bulstrode."[17] As a result, Delany's natural craftwork must be read within the larger, international context of botanical science as well as the domestic philosophies behind vegetality. Her crafts were thereby never just idle projects of a woman with free time and plenty of resources; instead, they represented the ways in which naturalist knowledge played a key role across multiple spheres of production in the eighteenth century. Just as historians have pointed to the botanical knowledge required to produce dyed cloth in the period, for example, significant botanical knowledge was necessary to produce the kind of expensive paper Delany preferred for her mosaics, sketches, and other collage work.[18] By attending not only to these kinds of material details but also to the emotional underpinnings of her craftwork, I highlight the ways in which naturalist study had significance to women on both personal and communal levels.

Collections and Reflections: Delany's Domestic Conchology

Like many naturalizing women in the eighteenth century, Delany valued the pursuit of knowledge as an intellectual goal that was desirable for a variety of reasons. As she advises her niece, "All human creatures have a natural inclination to knowledge, which is generally called curiosity, and if rightly directed it will be the means of great improvement to yourself."[19] A characterization of intellectual drive consonant with most measured assessments of women's academic pursuits (the emphasis on "rightly directed" inquiry being a crucial qualifier), this approbation of curiosity in women still provides a space for intellectual resistance against an increasingly conservative educational climate. The desire for knowledge—in this context, explicitly scientific knowledge—is a powerful tool for the enrichment of women's lives. Writing from the perspective of a widow with more intellectual and personal freedoms than her unmarried niece, Delany is nevertheless able to offer the same scholarly activity as consolation to both her niece and herself; the pursuit of natural history transcends social roles, age, and even class.

Along with several of her naturalizing predecessors, Delany found the true value of natural study to be tied inextricably to a profound sense of spirituality. A deeply pious woman, Delany often looked to her faith for guidance on personal issues as well as intellectual and creative pursuits.[20] Throughout her life she framed her interest in nature much as Charlotte

Lennox or Priscilla Wakefield did, with a combination of wonder and awe that encompassed both the natural world and the Being credited with creating that world. As a young woman, Delany gushed to her sister, "I have got a new madness. I am running wild after shells. This morning I have set my little collection of shells in nice order in my cabinet, and they look so beautiful, that I must by some means enlarge my stock; the beauties of *shells* are as *infinite as of flowers*, and to consider how they are inhabited enlarges a field of wonder that leads one insensibly to the great Director and Author of these wonders."[21] Like Lennox's injunction for her readers to allow the contemplation of nature to expand the mind outward, Delany conceptualizes her "little collection of shells" as the inspiration for a contemplation of the infinite divinity of creation. Further, while the "madness" of natural history and the "wild" nature of her appeal evoke the kind of sexual appetite that conservative commentators like the Reverend Richard Polwhele disparaged in intellectual pursuits like conchology and botany, such enthusiasm is balanced with an equally exuberant "wonder" of the "great Director and Author." For a young Delany, the pursuit of natural study is as exciting as a glimpse of Queen Caroline's gown, and as awe-inspiring as prayer. With beauties "as infinite" as the more easily accessible flowers, shells were and would remain both scientifically, aesthetically, and spiritually valuable to Delany throughout her career. As Amanda Vickery notes, her philosophical researches and her craftwork are best understood "as a strengthening of her faith," rather than activities pursued alongside it.[22]

This declaration of "madness" also reflects the popularity of conchology with women of a certain class. Scientific texts sold extremely well in the second half of the eighteenth century, and conchology texts were among the most expensive and the most popular. Best-selling texts like Emanuel Mendes da Costa's *The British Conchology* (1778) and Thomas Martyn's *The Universal Conchologist* (1784–89) were lushly illustrated masterpieces that catered to a widespread aristocratic interest in the subject. Containing detailed instructions for the proper processing of shell specimens—an extensive procedure that included precise instructions on how to kill and remove the creature inside, how to clean and sterilize the shell, and how to prepare it for preservation and display—these texts were invaluable consultation resources for amateur naturalists and virtuosi alike; given her enduring interest in the subject, it is possible that Delany herself owned some of these texts prior to her collaboration with the Duchess.

Furthermore, like botany, conchology was a source of great artistic inspiration for Delany. In addition to decorating common household items like chandeliers or frames, she made intricate grottoes lined with shells at Delville and Bulstrode: intimate spaces created through a harmonious meld of female collaboration and organic construction.[23] She was apparently quite talented at creating shapes and forms with inlaid shells; as she writes in 1751 of work she was completing at her home in Ireland to share with close friend Anne Donnellan, "I am filling a nitch with a mosaic of shells and the compliment paid it two days ago as, that *"it was very like Irish stich."*"[24] Throughout the spring and summer of 1770, she continued work on another shell-lined grotto at Bulstrode in conjunction with the Duchess, writing to her niece in July that "the cave goes on amain; a month more I think will complete it."[25] Lisa Moore highlights the intimacy of the grotto space as a venue for feminine companionship, a companionship that I want to emphasize as fundamentally rooted in the mutually rewarding pursuit of scientific study.[26] As she does with the exchange of crafts, Delany thus builds relationships with other women—as a part of the Bluestocking network and elsewhere—through the study and application of conchology and other naturalizing pursuits.

Conchology was part of her married life as well. In 1759 she writes excitedly to her sister Anne Dewes of a "bow window" she was in the process of decorating at Delville: "I am making festoons of shell-flowers, *chained up* with silver shells, which will look very well on the crimson ground."[27] In adorning a window in the home she shared with her second husband, Delany demonstrates that scientific labor was not incompatible with the domestic duties which accompanied a marriage. While many women artists and artisans ceased their creative productions after marriage (or devised ingenious ways of continuing such work), it seems evident that Delany's pursuit of natural history was a welcome activity in her marriage. Dean Delany was also interested in the natural world (more specifically, botany and horticulture), and there is every indication that he warmly supported his wife's creative endeavors. It is true that Delany certainly had more financial means as a genteel woman and time as a childless woman than some of her peers did, but her passion for naturalist craftwork was also seen by her contemporaries as appropriately feminine.

The particular focus on shells in Delany's writings also suggests an important feminine parallel that Haywood's characters likewise evoked with their studies of snails: as "inhabited" natural artifacts, shells functionally model the kind of domesticity that women in the eighteenth century themselves were seen to embody. Living beings that literally bear their homes upon their backs, "inhabited" shells were marked as metaphorically feminine objects in much the same way that flowering plants were viewed as feminine. Typifying something fecund, something protective or maternal, these inhabited shells even physically suggest the feminine: as Moore elaborates, "The erotic and feminine connotations of shells, with their salty smell, vaginal shape, and association with Aphrodite rising out of the sea-foam" underscores the intimate relationship between women, shells, and the female network of production and exchange in which the resultant shellwork circulated.[28] Thus, while men collected shells as much as women did (although no one paralleled the Duchess of Portland), the connection between women and shells possesses a greater symbolic significance within an eighteenth-century culture in which such metaphorical relationships between women and shells were familiar. Shells were, paradoxically, symbols of both domesticity/home/maternity and sex/desire/exoticism. Much like the competing cultural scripts about women's sexuality that wound throughout the eighteenth century, the linkages between women and shells were both consistent and consistently unstable.

Along with a thriving public interest in shell collecting, this association led to the publication of several works of conchology by women in the first decades of the next century, including Mary Roberts's field guide *The Conchologist's Companion* (1824) and Sarah Hoare's *Poems on Conchology and Botany* (1831). As the expansion of natural history as a prevalent field of inquiry provided opportunities for women to participate in both scientific spheres and the publishing industry, it also linked women more closely to their natural subjects. By the last decades of the eighteenth century and the earliest decades of the nineteenth, women were able to lay claim to authority as naturalists, albeit with some limitations. While few women were able to compete with voyaging botanists like Banks and Solander or the many other Linnaean disciples, women could and did assume a specific kind of authority over domestic subjects, both flora and fauna. As I argued throughout the first section of this book, the fact that women metaphorically and pedagogically allied themselves with domestic plants, for example, gave them an advantage when it came to generating texts and knowledge about indigenous biota. Likewise, snails, conchs, clams, and cockles, as creatures both metaphorically domestic and biologically indigenous, were ripe subjects for women's scientific expertise.

Further, as Delany's raptures over "beautiful" shells suggest, her collected objects are valuable for aesthetic and emotional reasons (and therefore may be duplicates), rather than solely because of their significance as unique representations of a type (which was more common with those assembling professional collections). In this way, her collecting practices are characterized by two convergent ideas: the personal significance of a natural object, and the scientific value of that object. Her shells were also often of personal significance because of their source. While the Duchess and Delany often went shell hunting at Weymouth (the estate of the Duchess's daughter Elizabeth), they also both acquired shells sent from friends across the country. Many of Delany's own specimens, for example, came from her family: in January 1778 she writes to her grand-niece, "Pray give my love to your brothers for the *curiosities* they have been so kind as to send me, and to your mama for the shells, some of which are the best of the kind I have met with."[29] One specimen, "that which was wrapp'd in the brown paper," even became incorporated into the Duchess's personal cabinet—not an easy feat, given her extensive network of resources and penchant for voracious shell collecting.[30]

This specific shell, moreover, reinscribes the important connections between conchology and domesticity: Delany writes to inform her grand-niece that the donated shell in question was a *Helix arbustorum* (today, *Arianta arbustorum*), a common land snail that was indigenous to England. Unfortunately, this snail does not possess a very noticeable or beautiful shell: a typical *Arianta arbustorum* is brown, with a few yellow spots or brown bands, adapted to allow the tiny snail to blend into its surroundings. As a diminutive shell that might have been found practically anywhere around St. James's Place or Bulstrode, this specimen is seemingly at first kept out of affection for a beloved young relative, rather than for its conchological significance. However, the fact that this small shell is both "the best of the kind" and a gift from a darling grand-niece reinforces the mutual relationship between science and female intimacy. Conchology is not only a source of intellectual authority for women but also a locus for the development of interpersonal connections: women collect together, dissect together, and craft together.

There is also evidence in Delany's correspondence to suggest that other women in her life besides the Duchess actively collected shells, including her grand-niece Georgina Port, whose youthful visits to her aunt were often spent on naturalist expeditions with Delany and the Duchess. Georgina also seemed to have some acuity with finding out choice or rare shells: in September 1780 Delany lovingly relates to her niece that the Duchess had created a special place of pride for "*Miss Port's shell*" among her vast collections.[31] In fact, Delany's young grand-niece played an active role in the conchological endeavors of both adult women. After a visit to Upton to see her friend Dr. Fothergill, a visit for which she brought along Georgina, Delany writes to her niece Mary (Georgina's mother) of the little girl's exuberant reaction to participating in botanical and conchological expeditions. She explains affectionately, "I took *my little bird* and Miss Pott to Upton in Essex, 10 mile off, to Dr. Fothergill's Garden . . . G.M.A. delighted, fluttering about like a newborn butterfly, first trying her wings, then examining and enjoying all the flowers . . . Thursday she was invited to drink tea at Lady Stamford's . . . [and when] I call'd for the precious charge . . . [she] brought away shells in abundance; and her collection encreases so fast that you must provide her with a cabinet to keep them."[32] During a visit to her great-aunt, Georgina gathers not only botanical but also conchological subjects, learning to replicate the same kind of naturalist work being undertaken by the adult women. The presumptive arrival of a specimen cabinet in Georgina's near future also reinforces the pedagogical value of naturalist labor. As a young girl under Delany's care, Georgina was consistently kept busy at various tasks, both domestic (sewing or mending) and scientific (sorting and collecting of natural specimens). The specimen cabinet represents a step forward in her pursuit of natural history as she moves from idle childhood collector to true amateur naturalist. Indeed, given that Delany and the Duchess also regularly sent Georgina a number of other natural specimens (as they had done for her mother, Mary), including a "gold pheasant's feather" from one of the Duchess's resident fowl, her specimen cabinet would fill quickly.[33] In this way, Delany and the Duchess practiced natural history and passed on the passion for naturalist pursuits to a younger generation: Lady Weymouth was already herself an enthusiastic conchologist. The practice of natural history becomes a multigenerational affair, an intellectual and emotional pursuit that unites women across boundaries of class, place, and time.

Delany's enthusiasm for naturalist collecting sustained her throughout her long life, as evidenced by an excerpt from a letter written during her second widowhood. While the excitement about her natural collections is tempered, the intimate affiliation between natural history and spirituality remains: "No amusement gives me such pleasure as my shells and fossils, and every acquisition, tho' ever so small, if good of its kind, adds to my pleasure. I hope it is not only the beauty and variety that delights me; as it is impossible to consider their wonderfull construction of form and colour, from the largest to the most minute, without admiration and adoration of the great Author of nature."[34] Delany's enduring interest in the gratifying potential of natural study is clear. The study of nature is both intellectually and spiritually fulfilling, as Delany encourages the curiosities of form and color alongside the marvel of their construction. The value of such contemplation is thus twofold, and evidently neither benefit was privileged over the other; in fact, the two seem inextricable from one another, as she declares it "impossible to consider [the specimens] . . . without admiration and adoration of the great Author of nature." It is also clear that part of her satisfaction with her dear collection of "shells and fossils" derives precisely from its status as a collection, in which the abundance of specimens evokes the wondrous appeal of the abundance of the natural world: the project of nature's "author" is replicated in miniature in her very own closet.

At times, Delany overtly identifies the study of natural history as an emotional comfort and a source of spiritual calm and serenity, hinting at the multiple benefits that amateur models of natural history could hold for their practitioners. On a day when she found herself at loose ends, without work to do, she tried to interact with the nonhuman denizens of Bulstrode, but found that the creatures of the Duchess's menagerie would not oblige. She relates to her grand-niece, "I hurried into the Garden. . . . I wanted my alluring companions, and they would not leave their retreats; they peeped through the branches of their sweet bowers. . . . I mused on till I came to my favourite bench, where we last sate, and there I invited all-composing contemplation."[35] Although one might read this passage in part as expressing the frustration of naturalist practice—after all, what is there to study when one's subjects refuse to emerge?—the study of natural history for women was often equally about the experience and the attempt to explore as it was about the acquisition of new objects, however thrilling such acquisition may be. Even without the benefit of specimens and samples, Delany found the connection with nature to be a spiritually uplifting one, a contemplation both intellectual and divine that formed the backbone of her own scientific philosophy. Surrounded by creatures and plants (albeit hidden ones), she found solace from the cares and worries of a long, often difficult life. Ultimately, this "all-composing" meditation defines her relationship with the natural world: between collecting, cataloging, and creating, reflection upon the multiple benefits of natural study percolated everything Delany did.

There was clearly a soothing aspect to the practice of conchology that complemented these moments of excitement and joy. In January 1772, a few short years after her second husband's death in 1768, Delany writes to her brother Lord Bernard Granville with a sundry list of gossipy updates and information about their mutual acquaintances that closes in a somewhat surprising manner: "I have no news; I am quiet in my chimney corner, and amuse myself with settling my shells that have been all jumbled together, and feel great comfort in having a place I can call *my own,* though it has cost me more than I intended."[36] Having moved to St. James's Place only after the death of her much-loved second husband, her comfortable solitude is bittersweet. However, one solace for the occasionally lonely Mary Delany is once again conchology, as sorting through her shells and devising a system of organization for their current "jumble" both amuses and relaxes her. The ability to look to naturalist practices for reasons beyond intellectual engagement or fulfillment seemed especially valuable to a widow like Delany. A later letter the same month references her (unaccomplished) plans to set aside the "morning for rummaging and settling fossils" in order to display them in the parlor, indicating that organizational tasks were as much a part of the naturalist's profession as the collecting expeditions.[37]

Her reference to fossils also reflects the breadth of Delany's scientific interests and the kinds of inquiry she conducted without processing the subject into any kind of craft. Used instead as objects of display in the home or in outdoor garden spaces, fossils were items liberally exchanged between estates, representative of various soil compositions, climate, and other regional differences that produced a diversity of stones and minerals. She was particularly fond of cataloging fossil specimens and often illustrated the texts herself: Llanover cites a manuscript containing "39 drawings of the crystalline forms of minerals."[38] Delany was especially proud of some samples sent from her niece, as she writes, "We also fixed upon a place in the Cave for the Ilam fossils, which are much admired, especially the great rock which is covered with coral, scientifically called *Madrepores.*"[39] Utilizing technical terminology alongside the more colloquial appellation of "coral," Delany's recording of her mineral collections captures equally her genuine intellectual pleasure and her contemplation of the natural processes that produced those minerals.

Several fossils also occupied enigmatic spaces for Delany's circle of philosophical friends. The above-mentioned "Ilam fossils," or fossilized fragments of shell, were found at her niece Mary Port's family estate at Ilam. In spite of having sent these fascinating subjects of study to her aunt, Port apparently provided insufficient information to Delany, as the latter writes imploringly, "You must inform us of their birth and parentage, particularly of some brown moss-like substance that was pack'd into the largest cockle, and a little brassish, coperish, goldish thread-like stuff adhering to a bit of slate or coal, and which has puzzled even Mr. Lightfoot."[40] These inadvertent hitchhikers on the uncleaned fossils—moss, bracken, or the like—confuse Delany and her companions as they perform the organizational and processing work left incomplete by Port. The extraneous flora and fauna are also exciting mysteries, as Delany eagerly writes to her niece to "inform us where they were found, whether on rock or tree, or bog? you must be very minute in your account; nothing less can satisfye such accurate enquirers."[41] Including herself in the "accurate enquirers" alongside the expert Lightfoot, Delany creates a balanced network of intellectual inquiry among women and men, neither deprivileging her own expertise nor too highly extolling Lightfoot's. The Ilam fossils thus serve multiple emotional and scientific purposes, as connectors between female family members and the inspiration for a shared scientific inquiry between experts and amateurs.

During this period of her life, Delany was living half the year with the Duchess of Portland, and many of her specimen acquisitions occurred in tandem with her dear companion. As indicated in an earlier passage by the caveat of "tho' ever so small," Delany was likely given second pick of any specimens sent to the Duchess for the latter's own, more extensive collection. Delany thanks her niece for the gift of a number of mineral samples: "The sparr (or rather chrystalization) and mundic in the coal is very curious and much admired. You know how pleasant it is to offer a mite to our dear, kind Duchess, and you may be sure I gave her the *best bits*."[42] However, it was common for the two women to share samples and specimens, whether discovered, gifted, or bequeathed; for example, a Mrs. Bristow explicitly left specimens to both the Duchess and Delany in her will.[43] The two women were closely allied through their scientific interests and created at Bulstrode a thriving scientific community that incorporated women from all over England, including Viscountess Elizabeth Andover, Elizabeth Montagu, Frances Boscawen, Lady Mary Bute, Lady Gower, Countess Georgiana Cowper, Lady Weymouth, and even Queen Charlotte herself.

In a well-known letter to her sister, Delany describes a typical day at the Duchess's estate, sketching a vision of "her Grace's breakfast-room, which is now the repository of sieves, pans, platters and filled with all the productions of *that nature*, are spread on tables, windows, chairs, which with books of all kinds, (opened in their useful places) make an agreeable confusion."[44] Here we see domestic spaces converted into scientific ones, as the breakfast room becomes the buzzing hub for the processing of shells, an activity conducted indoors more often than not, due to the fragility of the specimens and the unwieldiness of the "useful" reference books. Furthermore, domestic objects more ordinarily found in the kitchen—"sieves, pans, platters"—are repurposed for scientific tools in an active conchological survey. While certain critics have taken this notion of "agreeable confusion" to represent a chaotic scene of women's scientific pursuit gone wild, Maria Zytaruk more perceptively gleans the positive associations of such a scene; she notes, "For Delany, it was the unstructured domestic setting, where books, objects, and individuals jostle for space, that produced unfettered and fruitful conversation."[45] For these women, true scientific inquiry took place in a bustling, vibrant female community, as we imagine the Duchess examining one specimen, Delany consulting upon another, and servants washing untreated shells in acid to sterilize them or brushing off the pellicles (the organic skin that forms on the calcium of the shell) to brighten the samples for display.

The Duchess's expansive naturalist collections were praised during her life and after her death by expert men of science and amateurs across gendered and social lines. The auction catalog (which introduced the itemized contents of the Portland Museum after the Duchess's death in 1786) glowingly writes of her "diligent . . . Enquiries into Natural Knowledge and the Polite Arts" which led to the "Valuable Collection . . . she had with so much Pains and Treasure accumulated."[46] Indeed, her passionate and dedicated collecting over the course of her life inspired the following admiring speculation: "It may be remarked, that the celebrated *Linnaeus*, who had studied the Subject, and methodized the Materials of it, has not described One Fourth Part of the Objects contained in the Museum."[47] Ultimately, the catalog declares, "in this Branch of Nature's Works, no Collection in *Europe* can equal that of her Grace's in Number and Variety."[48]

While the Duchess provided the bulk of the material resources for the ladies' joint scientific endeavors—including but not limited to the processing space, specimens, reference texts, assistants, and tools—their joint investigations into the natural world highlighted the value of the Duchess's patronage to Delany's scientific and artistic endeavors. The former's emotional, material, and financial support enabled many of their shared scientific experiences and facilitated the active creative and intellectual labors of the latter. Delany notes proudly, "Botanic inquiries are never neglected" at Bulstrode, and nearly every day found the women dedicated "to plants, to books, and to work."[49] "Work" in this context likely meant both the conventional plain or fancy work of sewing and embroidery and the work involved in procuring, processing, and cataloging shell or plant samples. Much of Delany's correspondence is punctuated with this dual function of "work"; given the close allegiance between her interest in natural history and her domestic work, such ambiguity is useful in discerning her conceptualization of the role that natural history played in her day-to-day life. And, in spite of their fundamentally unequal relationship, it seems that both women collaborated within a broader network of scientific exchange that expanded far beyond the borders of the Duchess's lavish estate.

Delany frequently self-identifies among the professionals or "virtuosos," creating a space for herself in the community of renowned botanists, explorers, and naturalists that circulated in and out of Bulstrode. She writes to one correspondent, "To-night or to-morrow we shall be *illuminated* by the two celebrated philosophers—Mr. Pennant and Mr. Lightfoot, and *virtû* will be in its full glory."[50] While the presence of male naturalists added to the value of their scientific community, it did not entirely define their scientific activities or compromise the scientific integrity of their pursuits. Rather, visitors were viewed as stimulating opportunities for the ladies to share their discoveries in a mutually rewarding discussion. As Delany excitedly reports on one occasion, "Dr. Solander, &c. came as expected, and I am now going to get a botanical lecture and to copy a beautifull flower called the *Stuartia.*"[51] In this scenario, both Delany and the lecturing Solander are conducting scientific work: as he reports on the basics of botanical research or recounts stories from his South Sea voyages, she works on one of her own mosaics, a blossom from the Stewartia Malacodendron—an exciting, newly imported exotic specimen. Here, her botanical work is as cutting-edge, if not more so, than anything Solander may have been lecturing on. As Lisa Ford notes, "Delany was representing newly introduced exotics [in her paper mosaics] within three or four years of the first date of introduction," emphasizing again her intimate access to the most recent botanical discoveries.[52] Crucially, this elite access to botanical resources also transcended her friendship with the Duchess; as Ford has meticulously traced through an accounting of Delany's letters and the provenance notes on the mosaics, the artist often produced mosaics at the estates of friends and family members with whom the Duchess had little personal connection.[53]

Delany also evidently had a reputation among friends for a tireless pursuit of natural history. Frances Boscawen evaluates Delany's work as more intellectually significant than the conventional business of upper-class women, as she concludes a letter, "So much for my journey and occupations. Yours, my dear madam, thro' the paths of science are far otherwise important."[54] This qualification from a peer seems crucial; although many of her contemporaries also employed natural motifs in needlework or craftworks, Delany's productions had achieved among her fellows the reputation of "science." While the word "science" in the eighteenth century usually meant any kind of profound knowledge, here it refers clearly to Delany's interests in botany, conchology, and entomology and the resultant creative productions.

Like the Duchess, Delany also seems to have been consulted as an expert in identifying plants. She dictates to her niece, "I am sorry you did not send your little flower, it may come yet; I believe by your description it is a *linaria* or *toad flax*."[55] Llanover even writes of a botanical treatise that Delany may have produced: "On the 18th October of this year (1769), Mrs. Delany commenced a manuscript book at Bulstrode, which she completed with her own hand. It appears to be a translation of the first edition of Hudson's '*Flora Anglica*,' published in London in 1762."[56] Llanover speculates that it may have been a copy or imitation of Lightfoot's manuscript notes for *Flora Scotica*, which seems more plausible, given that the latter was written in English. The fact Delany's scientific knowledge was sufficient to annotate and organize the manuscript also suggests to a modern audience just how extensive her naturalist background really was. Some of the more remarkable things about this manuscript, moreover, as Llanover notes, are the "appended list of the *Genera* in Latin" and the personal notes about particular specimens. For example, Delany makes a small note near the entry for the "*Fir-coned Hydnum*" (today, *Auriscalpium vulgare*, or pinecone mushroom) that reads, "This was found at Bulstrode on fir-cones, in November 1769."[57] A species first described by Linnaeus in *Species Plantarum* (1753), this fungus was not unique to England, but it had not previously been part of the typical record of indigenous mushrooms, making its sighting at Bulstrode a noteworthy event. Thus, Delany was not only working intently with texts about indigenous flora; she was also actively pursuing expeditions to seek out such specimens.

In fact, it seems as though it was assumed that she would be at botanical work nearly constantly, as Delany writes wryly to a companion, "Most likely you at this instant imagine me gaping and staring at all the wonders in the Oxford Museum, or hobbling upon the pebbles from college to college, or winking and blinking in the burning sun over the plants in the phisick garden."[58] While these activities were not out of the ordinary for a woman of her means and interests, this particular itinerary suggests a greater dedication to vigorous scientific activity than a contemporary might expect of a woman. The letter, however, was written from Bulstrode, which was located some distance from an urban center; Delany admits to the Viscountess Andover, "No, here I am enjoying every rural delight (which abound here), and gratifying every sense."[59] In spite of her well-deserved reputation as an active naturalist, Delany also enjoyed an emotional connection with nature that didn't need to involve the humorous "gaping," "winking," or "hobbling." Instead, like her visit to the Bulstrode menagerie, this time spent in nature without an active scientific agenda allowed Delany to continue to cultivate her intimate relationship with nature.

Yet to spend hours poring over the plants in the Chelsea Physic Garden was commonplace for Delany, as were ventures throughout the grounds of Bulstrode, around the Royal Botanic Gardens at Kew, and to the estates of friends like Lord and Lady Bute's Luton Park and Lady Gower's Bill Hill.[60] Natural history craftwork and the collection of scientific specimens were vital parts of a fulfilling life for Delany and the Duchess, as engagement with scientific subjects structured each of their days and often dictated many of their social activities. From collecting parties in Dover with the Duchess to botanical lectures at Bulstrode to diverse projects involving shells, petals, fossils, and feathers, the natural world played a role in nearly everything Delany did. Furthermore, she undertook each activity with a thoughtful approach that sought to negotiate space for an amateur practice of science within an increasingly professionalized botanical world. Flourishing at the convergence of science and art, Delany's craftwork invaluably contributed to a vibrant new discourse of botanical study that circulated among Bluestocking women at the end of the eighteenth century.

The Names of the Plants

While many of Delany's craft productions involved shells, feathers, fossils, and other natural objects, her most famous work featured botanical subjects. To engage with botany in the latter third of the eighteenth century was, inevitably, to enter into the debate about Linnaean taxonomy. While the Linnaean taxonomic scheme was controversial for a number of reasons, it remained the dominant model for botanical study in the last third of the eighteenth century. The system was easily learned by educated men of science and amateur botanists alike, as Patricia Fara explains: "Determined to make his ideas easily accessible, Linnaeus wrote clearly, ensured that his books were cheap, and gave practical instructions about collecting, labeling and growing plants."[61] His popular *Philosophica Botanica* (1751) "ends with a series of one-page instructions, teaching the reader how to set up a herbarium, organize an excursion, plant a garden, and even embark on a voyage of discovery."[62] Both relying upon and contributing to the rising botanomania of the late eighteenth century, Linnaeus's work was tremendously influential—particularly for female authors who seized upon the creative possibilities of such a system. Later in the century, authors such as Maria Jacson (whose *Botanical Dialogues* actively and almost exhaustively covers the details of Linnaean classification, as detailed in chapter 2) and Frances Arabella Rowden (whose *A Poetical Introduction to Botany* featured lively botanical poetry depicting each class in the Linnaean system) took advantage of the popularity of Linnaean botany to express their own scientific interests for female readers and young audiences of both genders.

Like Rowden, Delany strategically adapted the Linnaean taxonomic system to her creative project in ways unanticipated by its originator. The decision to use Linnaean taxonomy as a means of identifying her mosaics was at least partially one of convenience; in the absence of another clear alternative, adhering to dominant classificatory convention seemed logical. The painter Georg Ehret, who influenced Delany's botanical development, actually used no Linnaean terminology in his own illustrations unless asked to—a decision that likely arose from Linnaeus's use of Ehret's illustrations without permission or credit.[63] Perhaps inspired by Ehret, Delany's use of Linnaean concepts in her mosaics is inconsistent. A typical mosaic—if there can be said to be such a thing—bore the initials "MD" in cut paper in a corner on the bottom of the front side, plus a small white label featuring Latin and English names. The reverse of the mosaic bears handwritten information about where the mosaic was completed, by whom the plant specimen had been donated to Delany, and the Linnaean categorization of the plant featured on the front. This was hardly the case for all of the mosaics, however: only about half featured the Linnaean categorization, while most of them did include the Latin name for the plant; a small percentage of them included the plant donor (it was most often Kew or Chelsea Physic Garden); and a larger percentage of the mosaics indicated where the mosaic was produced (usually at Bulstrode, although also frequently at St. James's Place).

Delany, who was otherwise an excellent writer and correspondent, was erratic in her formatting of the Linnaean classification system, writing, for example, "Syngenesia Poly: Super" for the *Tulsilago alba* in 1777, but then "Syng. Polygam. Superfl." for the Erigeron in 1778. Likewise, the former is labelled with the vulgar name "Colts Foot or White Butter-bur," while the Erigeron has no commonplace name (it usually refers to members of the daisy family). She also occasionally spelled the Linnaean terminology differently, such as labeling the *Indigo fera* as "Diadelphia Decandria" in 1777, after having marked the *Robinia pseudacacia* as "Dydelphia Decandria" in 1776. Her Latin names were also sometimes idiosyncratic terms—many of which are no longer accepted names for plants today—that she received from Solander or her other acquaintances. While this level of inconsistency might at first suggest a lack of botanical knowledge or a failure to maintain a rigorous labeling system over time, given Delany's stellar reputation as a writer and the persistently methodical nature of the floral portion of the mosaics themselves, it becomes difficult to sustain such an assumption. Since there is no way of knowing how purposeful such decisions were, the most we can gather is that rigid adherence to Linnaean taxonomic systems was less important to her than the vibrant artistic representation of the plant itself. Taxonomy, for Delany, seemed to be less about the epistemological value of naming a plant and more about the practical notion of being able to label that plant for an outside audience.

Similarly, by only partially applying the Linnaean binomial to her artistic work, Delany was able to exploit the limits and the creative potential of the taxonomic system. As Moore has argued, Delany's mosaics were particularly expressive in their adaptation of the Linnaean model and "can be read as explorations of both female intimacy and female sexual anatomy."[64] I would further add that through her adaptation of the fundamental association between women, plants, and sexuality that was concretized through post-Linnaean botanizing culture, Delany's aesthetic decisions in the construction of the mosaics emerged as an alternative means of presenting botanical knowledge. If femininity and plants were inextricable from one another in the eighteenth-century cultural estimation, then Delany's artwork celebrated this fact through the vibrancy and realism of her depictions. Erasmus Darwin, for one, seemed to recognize the linkage between Delany's work and her powerful evocation of the sexual system in plants, as he extolls her artistic virtues in *The Loves of the Plants:*

> So now DELANY forms her mimic bowers,
> Her paper foliage, and her silken flowers;
> Her virgin train the tender scissars ply,
> Vein the green leaf, the purple petal dye.[65]

The emphasis on Delany's purity alongside the artistic power of her "mimic bowers" is remarkable, a compliment that was due to Delany's own impeccable reputation and intended as a sincere homage to such a renowned creative force. Later lines characterize her creations as so life-like, that they even trick a personified Winter, as he "smooths his stern brow, delays his hoary wing, / And eyes with wonder all the blooms of spring."[66] For Darwin, Delany's creations are so vivid, so evocative of the true, vibrant natures of plants, that she has transcended the static qualities of artwork. Lauded by one of the strongest proponents of the literary and artistic potential of Linnaean sexual system, Delany's mosaics are clearly much more than realistic or charming: they possess a compelling power of their own.

Darwin's verse—his response to Delany's craftwork—also evokes for the modern reader the kind of distinction between and among individual plant specimens that twenty-first-century botanists have articulated as key to thinking about plants as species with their own needs, desires, and agency.[67] As Darwin himself explored in *Phytologia* (1800), plant behaviors and reactions to various sensations indicated the presence of some kind of central nervous system, thereby implying an intelligence or even personality among plants. While Delany gave no indication of being invested in the possibility of plant intelligence, it was common for her to express her interactions with individual specimens in emotional terms. For example, in June 1776 she writes, "I have made an appointment for to-morrow with a very fair lady called 'Lychnidea.' If I neglect her, she will shut herself up, and I shall see her no more."[68] Evoking a soft sensuality, Delany's description of the Lychnidea personifies the plant in multiple ways. Overtly demarcated as female ("a very fair lady") and then assigned feminine behavioral traits such as shyness, coyness, or even sensitivity in order to describe the plant's limited flowering period, the Lychnidea becomes an alluring woman even prior to her portrait. The flower is here no longer a mere scientific specimen, but a character in a broader artistic narrative. A postscript to the same letter depicts the encounter almost like an assignation, as Delany notes with satisfaction, "The fair lady was true to her appointment, and we parted friends."[69]

In assigning plants personality—needs, wants, decisions—Delany is doing more than cheerfully applying common gendered metaphors to flowering plants. While she is, like Darwin and many other authors, anthropomorphizing the Lychnidea, she is also relating in personal ways to her individual plant specimen. Lychnidea, or, to use the Linnaean term, "phlox," was indigenous to North America and temperate portions of

Asia. It had also been artificially introduced to Austria, making it possible that Delany received her specimen either from a European connection or, indirectly, from Asian voyages. It was first referenced in *Species Plantarum*, but was relatively new to England at the time she produced her mosaic. For Delany, then, "the fair lady" was an exciting acquaintance—and an acquaintance with a limited life span. Working with apparently only one cut specimen, Delany does not have repeated access to this rare plant. If she neglects to copy the phlox within a short period of time, it will wither and die before she has the opportunity to transform the organic specimen into material art. The sad reminder "I shall see her no more" evokes not only the transience of all uprooted plants, but the particular lifespan of this individual plant. This phlox, a plant for which Delany prefers the feminine-sounding term "Lychnidea," is one among many in acclimatization gardens throughout Europe—certainly one among untold numbers in the world, but still only one singular plant.

There is only one specimen labeled "Lychnidea" in the entirety of her mosaic collection, the *Phlox undulata*, which is dated August 11, 1781—more than five years after Delany mentions her "appointment" in correspondence with her niece. Of her 1776 project, Llanover includes a footnote in her edited collection of Delany's correspondence that states, "Mrs. Delany made a beautiful representation of a white and lilac lychnidea from nature."[70] A mosaic completed in 1776 entitled "Phlox Suaveolens" seems the most likely candidate, but the colors do not seem consistent with Llanover's descriptions. Beyond assuming that this is the "beautiful representation" Llanover refers to, we do not know why this specimen was "Lychnidea" in correspondence but the more technical-sounding "Phlox Suaveolens" as the final product. We do know, however, that this particular specimen became contextualized and particularized in very specific ways. For one, Delany's strategic use of the name "Lychnidea" reflects a negotiation with contemporary standards of botanical knowledge: the term was purportedly coined in Leonard Plukenet's *Phytographia* (1691–92) and subsequently used in Johann Jacob Dillenius's *Hortus Elthamensis* (1732). It was the common name for the flower until, according to the possibly apocryphal Elizabeth Washington Wirt's *Flora's Dictionary* (1837), Linnaeus misunderstood the classical reference and adopted instead the Greek work for "flame."[71] By evoking in her correspondence the traditional name for the plant, Delany signals a departure from the often-technical approach she took to categorizing her mosaics. Instead, she pursues a personalized, feminized relationship with this plant, articulated around a time when she held the plant in her hand, inhaled its fragrance, and noticed its texture. The personal relationship with her plant specimens weaves its ways throughout the entire collection of mosaics, known collectively as the *Flora Delanica*.

Botanical Artwork and the *Flora Delanica*

Mary Delany's work has usually been distinguished from that of other botanical artists in the eighteenth century because of her seemingly haphazard and enthusiastic pursuit of multiple artistic and craftwork genres. Further, her most well-known creative production, the floral mosaics, formally and visually defied artistic convention: they did not constitute typical flower paintings, but neither did they fit within the usual definitions of botanical illustration. Botanical illustration and flower painting, though similar in production, were two artistic schools that functioned with distinct yet overlapping conventions in the depiction of flowers. Flower painting had more flexibility in its artistic conventions and often included fanciful contextual detail such as surrounding plants, insects, or even broader geographic elements like streams or fields. Such paintings, like those in Robert Thornton's *Temple of Flora* (1799–1807), were not generally considered scientific. Women were often considered adept at this kind of artwork, including such notable British figures as Margaret Meen, Mary Lawrence, and Mary Moser (not to mention the many iconic Dutch artists of the period), but these artists were classed as flower painters.

Indeed, despite the fact that the painting of flowers either in landscapes or as a bouquet was a frequently cultivated accomplishment among women of a certain demographic, women were rarely categorized as botanical artists in the eighteenth century (the rare few including Elizabeth Blackwell and Henrietta Maria Moriarty, the subjects of chapter 4). Like most other areas of scientific study in the period, botanical artwork was considered a masculine prerogative, as it depended upon a significant knowledge of botanical minutiae and, later, of taxonomy. Further, like the study of botany itself, floral depictions in the period could be interpreted in a variety of ways; depending upon the position of the blossom, the stems, the leaves, and the reproductive organs of the plant, the portrait could be decorative, informative, or even provocative. To produce work that was recognizably botanical, rather than floral, in nature required an adherence to certain key standards.

Although they shared many similarities by virtue of sharing the same subject, botanical artwork influenced by Linnaean standards and common flower painting diverged on a number of key attributes. While depictions of plants in a pre-Linnaean artistic world (as well as those illustrations disinterested in serving a classificatory function, like the ones in a florilegium) often included a full, rich depiction of the plant from root and bulb to leaf, stem, and flower, true botanical illustration from the 1740s onward generally incorporated a few central elements: "*Trompe l'oeil* detail, [with] each stamen and pistil distinct, and the flowers themselves ranging from buds to fully-mature blooms."[72] Botanical illustrations in the second half of the eighteenth century depicted what is known as the ideal plant, or a temporally impossible illustration that "combines in the one image different stages of flowering and fruiting that would not occur concurrently but are essential parts of its identity."[73] Given the new sexual idea of plant identity, nonessential components in botanical illustration included most leaves, stems, and other nonflowering parts of the plant (the same components that were nonessential in the Linnaean taxonomic scheme).

Additional conventions included the use of a white or neutral background and occasionally dissected elements of the same plant positioned next to the blossom itself such as petals or sexual organs. After all, botanical illustration was about portraying a representative image of the plant that could be used for identification purposes, not about utilizing aesthetic conventions that might obscure such identification (or conventions that were irrelevant to its achievement). Due to its simplicity and botanical specificity, Delany's work does not accord with conventional interpretations of flower painting, but more closely fits within the sphere of botanical illustration. Yet her decision not to adhere to the notion of the "ideal plant" and to instead pursue a kind of detailed realism in her depictions (including wilted leaves or bruising) also distinguished her mosaics from typical botanical illustration. Thus, Delany's work occupied a space between these two modes of botanical depiction.

Most botanical artists, whether they produced mass-market magazines like Curtis's *Botanical Magazine*, expensive treatises, or one-off prints, also organized their collection of specimens in some fashion: geographically, for example, as Mark Catesby usually did, or by variety, as with Pierre-Joseph Redouté's *Les Roses* and *Les Liliacées*. Given the sheer breadth of Delany's samples, however, it is likely that her sample selection was motivated primarily by novelty and availability (with the exceptions, of course, of the lily for Queen Charlotte and the *Portlandia grandiflora* for its namesake, the Duchess of Portland). The fact that her model specimens were an impressive combination of acclimatized exotics and native plants is a testament to her expansive botanical networks, including the commercial Hammersmith nursery, the robust personal collection of Lord Bute at his estate Luton Hoo, and the expressly scientific royal gardens at Kew. Furthermore, as Ford has suggested, the varied provenance of her floral subjects allows us to read the *Flora Delanica* as a "[project] of remembering and keeping close one's circle of friends and acquaintances."[74] The mosaics merge multiple types of emotional intimacy—with friends, family, and plants.

The individual value of a given plant specimen, for Delany, was also tied to its unfamiliarity, and by the end of the 1770s she was intent upon depicting as many plant specimens as possible; in April 1779 she declares excitedly, "I am so busy now with *rare* specimens from all my botanical friends."[75] She was so delighted with the variety of exotic samples that Kew was providing her that she wrote proudly that "I am so *plentifully* supplied with the hothouse here, and from the Queen's garden at Kew, that natural plants have been a good deal laid aside this year, for foreigners, but not less in [favor]."[76] On another occasion, she relates of a trip with her grand-niece, "We return'd loaded with the spoyls of the Botanical Garden."[77] Treating nearly equally with "foreigners" (exotic plants) and plants indigenous to the British Isles, Delany created a collection of some emotional significance that also served as a comprehensive record of her access to plant specimens, and, therefore, of her lived experience as a botanist. Here I also note that, unlike her male compatriots, who named their botanical texts after the region of botanical focus (such as Curtis's *Flora Londinensis* or Lightfoot's *Flora Scotica*), Delany's project bore the name of its creator. Distinguished from competing and companion publications on multiple levels, the *Flora Delanica* was a truly unique production.

Delany's 1772 letter to her niece Mary Port with which I began this chapter provides very little detail about the origin story of the famed *Flora Delanica*. She writes only, "I have invented a new way of imitating flowers. I'll send you next time I write one for a sample."[78] Llanover provides a fuller description with an explanation likely derived from family lore:

> Having a piece of Chinese paper on the table of bright scarlet, a geranium caught her eye of a similar colour, and taking her scissors she amused herself with cutting out each flower, by her eye, in the paper which resembled its hue; she laid the paper petals on a black ground, and was so pleased with the effect that she proceeded to cut out the

> calyx, stalks and leaves in shades of green, and pasted them down; and after she had completed a sprig of geranium in this way, the Duchess of Portland came in and exclaimed, "What are you doing with the geranium?" having taken the paper imitation for the real flower. Mrs. Delany answered, that "if the Duchess really thought it so like the original, that a new work was begun from that moment."[79]

While she would later experiment with varieties of paper and methods of cutting, the basic procedure for creating each individual mosaic remained essentially the same. Delany layered cut paper and small amounts of paint onto a dark background—usually on pricey Chinese paper—in imitation of the live plant specimen before her. Although she was essentially producing what we might call a collage or a kind of decoupage, Delany would come to prefer the term "mosaic" for her original creations, due to their piecemeal composition.

In terms of technique and materials, the mosaics were unlike any other botanical depiction on record; John Edmonson acknowledges that he "know[s] of no precedent for creating such a faithful record using colored tissue paper."[80] Ruth Hayden notes that Delany had been talented with cutting paper even as a small child, and cutting out silhouettes without tracing would have been a skill used in other areas of domestic craftwork.[81] Kim Sloan further explains that Delany "had cut paper throughout her life, making not only landscape and figural silhouettes, but also three-dimensional pictures of birds, cut from vellum."[82] Delany may have dissected one sample of a plant before modeling another in order to aid her botanical specificity; she describes Ehret using a microscope to examine his floral subjects before painting them, and at Bulstrode she would have had access to such technology.[83] Further, her decision to create a collage instead of using paint was certainly not one of necessity; she painted regularly throughout her life, although less frequently as her eyesight diminished in her old age.

Llanover's proud description of Delany's paper mosaic project highlights one of the most frequently bestowed compliments that the creator was to receive throughout the remainder of her life: that is, the truly convincing realism of their depiction. Botanical artist and landscape designer William Gilpin noted that "she has executed a great number of Plants and Flowers, both natives and exotics, not only with exact delineation, and almost in their full lustre of colour, but in great taste. And what is the most extraordinary, her only materials are bits of paper of different colours. . . . These flowers have both the beauty of Painting, and the exactness of Botany."[84] This overlap between painting and botany, articulated explicitly by Gilpin, highlights the extent to which Delany's labor as an artist was reinforced by and founded upon her labor as a naturalist. This emphasis on "exactness," moreover, reflects the truly impressive achievement that Delany had earned through her mosaics—the reputation of a woman with scientific bona fides. While craftwork could be exact and often represented nature precisely (recall the embroidered screen inherited by Llanover), the fact that Delany's mosaics were seen as functioning within a different cultural sphere of production seemed to change the way in which they were evaluated. As botanical artwork, rather than craftwork, "exactness" mattered in a way that it had not previously.

Llanover reports a similar series of compliments from both artists and botanists, claiming that Delany's mosaics were

> admitted by Sir Joshua Reynolds and all the best judges of the art of drawing and painting to be unrivalled in perfection of outline, delicacy of cutting, accuracy of shading and perspective, and harmony and brilliancy of colours; while at the same time they were the admiration of botanists such as Sir Joseph Banks, Dr. Solander, &c. &c. Indeed, Sir Joseph Banks used to say that Mrs. Delany's representations of flowers, "were the only imitations of nature that he had ever seen, from which he could venture to describe botanically any plant without the least fear of committing an error."[85]

To some extent, Banks's approbation suggests that he interpreted Delany's mosaics as themselves a kind of botanical specimen, identifiable in the same way that live plants were, an important qualification to the notion of her mosaics as strictly art objects. While some modern critics have been oddly careful to emphasize that Delany's mosaics were not, in fact, technically accurate, the significance of these historical reports lies rather in the fact that her contemporaries considered the work to function in a scientific as well as an aesthetic sphere.[86] Their high approval ratings in a mixed population also importantly suggests that a product that was functionally considered craftwork—and therefore a more feminine object—could be and was appreciated in a way that transcended such limitations. This speaks to the influence of Delany's mosaics both within her inner circle of companions and with regard to the broader English public. The royal family made a point of reviewing Delany's progress on the project on several different occasions, inviting her to the Queen's Lodge as well as viewing the collection at Bulstrode.[87] Friends frequently wrote to inquire about her progress, even cautioning her against overwork: "I hope you do not *work* too hard, tempted by curious materials and *fine originals*, but rather you enjoy (with some degree of indolence) this tardy summer."[88] A distant and none-too-liked acquaintance even manufactured a pretense to visit Delany just to see her mosaics, much to the latter's bemusement.[89]

A certain degree of fame notwithstanding, her reputation for artistic and scientific ingenuity was well deserved. Delany founded her distinctive approach to botanical illustration upon her decision to use easily accessible materials and innovative techniques in order to create what was essentially a brand-new form of scientific artwork, a genre that distinguished itself from contemporary botanical illustrations in a number of ways. The most prominent botanical illustrator of the period, her friend Ehret, "always preferred vellum to paper, and body-colour to transparent wash, for his most important work; and he not infrequently painted his smaller plants rather above life-size to enhance their impressiveness."[90] Contrarily, Delany used very small amounts of paint on each mosaic to modify the paper slivers once they were placed upon the black background—a decision that a nineteenth-century reviewer of the collection viewed skeptically, noting that "it would have been far easier to shade a fragile stem with a little sepia, and equally effective to add a few high lights with a camel-hair brush, or to pen in the thin threads of the stamen."[91] Instead, Delany usually took advantage of the particular texture or pigmentation irregularities in her paper samples to evoke shading and texture in the mosaics themselves. Her plants were also usually replicated in comparable (if not exact) size, rather than to scale, with the frequent appearance of tiny plants and flowers on a larger black background making for an uneven number of proportionate compositions within the collection as a whole. Of course, she did modify the size of the black background depending upon the size of the floral subject; the *Flora Delanica* resists even proportions in a number of ways and in person evokes more of the scrapbook than the illustrated treatise.

By innovatively manipulating a limited set of materials, Delany demonstrates a remarkable inventiveness that surpassed that of her contemporaries and even those who would follow in her footsteps. Booth Gray, an acquaintance who created 136 collages of his own after Delany's method, traced the flower parts before cutting and liberally used paint in lieu of additional paper pieces.[92] The choice of a black background was also unconventional, as botanical artists typically preferred light or neutral backgrounds to soften the blossom's focus. In this, Delany may have been inspired by the Italian pietra dura, a stonework mosaic typically incorporating floral motifs set within a black background, or, perhaps, paintings by German botanical artist Barbara Dietzsch, who occasionally also worked on dark backgrounds.[93] A scant few of the mosaics were produced on light-colored or white backgrounds—one singular background is mauve, the *Annona triloba* in 1775—but they are far outnumbered by the striking black compositions.

In her creation of the mosaics, Delany also occasionally included small samples of the plant she was modeling in the mosaic itself, usually in a minor area of the composition. This incorporation of a plant fragment was sometimes found in other examples of eighteenth-century botanical artwork, but I argue that it carries distinct signification for a female artist, as I will elaborate below. Perhaps as a point of comparison between her version and the original, or possibly as a spark of inspiration, the function of the plant piece seems ambiguous. A survey of the mosaics yields the observation that a little more than 15 percent of the entire collection includes plant matter, usually referring to leaves, although in two instances it refers to petals and three cases it also includes what the British Museum has identified merely as plant fibers.[94] This inclusion of the organic specimen as part of the final crafted object not only reaffirms the exceptional cultural position occupied by these mosaics, but also reinscribes Delany's own investment in the critical and artistic possibilities of plant life. In a frequently reprinted review of the *Memoirs of the Life of Mrs. Mary Delany* (published by Biographica Britannica after her death), the author enthused that her mosaics "might justly be called a *forgery* of Nature's works."[95] This observation was intended as an effusive compliment, but the concept of falsely replicating an organic object raises some interesting questions about how Delany's craftwork was understood by her peers. Given that all botanical artistry is an artificial replication of an organic subject, how might Delany's mosaics differ in ways other than their unusual format? I suggest that the significant point of distinction for their reviewer—and for a modern audience—is the incorporation of actual organic elements within the composition itself. As a woman, already tied to nature and to flowers by countless literary and cultural references, Delany occupied a distinctive position with respect to these erstwhile plant pieces. By grafting onto the page real remnants of the flowers she chose as subjects, she produced a hybrid work of art that is part artificial (the paint and pencil) and part authentic (the leaf or petal). The fact that this hybridity translates as a "forgery," according to one review, reinforces the compelling intersection between craft and science that lay at the heart of Delany's artistic productions: her work is so good, it can fool even nature (as it has already fooled Winter for Darwin).

Delany did not set out to falsify natural objects, the Duchess's exclamation about the geranium notwithstanding, but her achievement of this uneasy realism recalls her personal relationships with her plant subjects. Created by and from plant specimens that were gifted to her from friends across England, these mosaics represent an intimate layering of her friendships with people and her connections with plants. Just as many of the mosaics were labeled on the back with the name of the plant donor, the mosaic on the front of the paper incorporated that very plant into its composition, thereby telling a story about a particular plant in a particular time and place. Freya Gowrley has discussed the narrative potential inherent in crafted material objects produced by women, and I want to build upon her astute analysis by linking the mosaic to these other types of scientific craftwork.[96] Where Gowrley argues for the crafted souvenir from the Grand Tour as an object that "signifies a relationship between material culture and the construction of personal narrative," I propose that Delany's plant specimens served as souvenirs of another kind, mementos of friends and family members that became embedded into her material productions. One particularly detailed mosaic, the *Selinum palustre* or *Marsh selinum*, is attributed to a specimen donated by Lady Weymouth. This mosaic (fig. 6) contains more organic matter than most of the other pieces, with nearly 50 percent of the composition taken up by a real stem and leaves. Creating a striking visual effect for a modern viewer—half of the composition is the brown of a dried plant specimen and the other half is comprised of vivid greens and pale purples—this melding of components also reminds us of the significant interplay between the material productions of women, their intimate relationships with one another, and their equally intimate relationships with plants. Thus, as these organic objects become both figuratively and literally part of her mosaics, they create interwoven narratives about people, places, things, and plants.

The plants that inspired (and, in some cases, formed the foundation of) these mosaics did indeed come from all across England and all over the world. The majority of her plant samples are not provenanced; however, about 10 percent of the specimens were recorded as coming from the Royal Botanic Gardens at Kew, and around 2 percent were attributed to Chelsea Physic Garden. The remainder of those with listed provenance

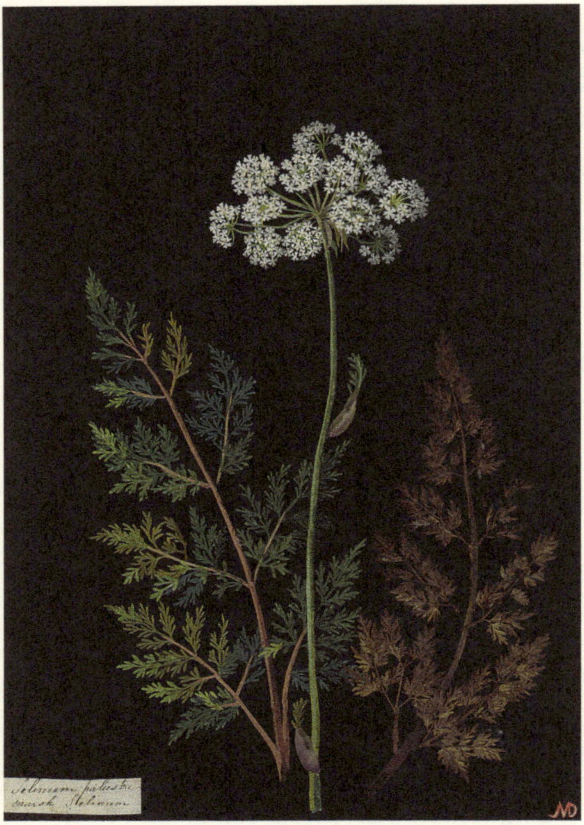

FIGURE 6. *"Selinum palustre,"* Mary Delany, 1775. (© The Trustees of the British Museum)

came from the estates of friends, ranging from the Duke of Northumberland to a Miss Jennings. Delany received more than forty from Hammersmith nurseryman James Lee, twenty-five from Royal Society Member Dr. David Pitcairn, six from Dr. John Fothergill (who cultivated his own very impressive botanical garden at Upton), and fourteen from her friend and fellow naturalist Lord Bute, among many others. What this brief survey of the plant sources suggests is that Delany's project attracted and benefited from broader public attention to her craftwork. As a de facto member of the Duchess of Portland's household, she may have had access to places like Kew regardless; the sheer breadth of sources, however, indicates that her mosaics engaged the imagination of gardeners and natural historians from across the country. This also connects her to a larger network of naturalists, from professionals like James Lee to amateurs like Lord Bute, who were all invested in multiple ways in her success as a botanical artist. Whether that investment was tangible and financial—many of the exotic plants would have been expensive to maintain—or emotional, Delany thrived as a result of their support.

While such support was evidently provided by naturalists and friends of both genders, her most ardent supporters were her female friends. Most of the mosaics not completed at her own home in St. James's Place were produced at the estates of female friends: the Duchess of Portland, Lady Gower, Mrs. Boscawen, and Lady Bute (in conjunction with her botanizing husband). In the extant correspondence, detailed suggestions for new plants usually come from women (although, as demonstrated, male acquaintances certainly sent her numerous samples): Frances Boscawen was especially excited to offer Delany the opportunity to use the "Chiramoya," or the *Annola cherimola* (custard apple), as the inspiration for one of her mosaics.[97] Boscawen includes a history of the plant with a letter written ca. November 1779 and declares proudly, "I think it may be in time promoted to a place in certain invaluable folios."[98] Furthermore, many of the samples Delany gave as gifts were to female friends and acquaintances; in addition to the lily done for Queen Charlotte and the *Portlandia grandiflora* for the Duchess, she also sent a mosaic to Anne Viney in October 1779.[99] The final collection itself was even preserved and eventually donated to the British Museum by a female descendent of a collateral line, Lady Augusta Llanover herself. Delany's work on the mosaics was thus shaped by her relationships with other women, just like her craftwork. For her, vibrant female friendships provided the support, material resources, and opportunities to engage in scientific study at a time where such a pursuit was not open to all women.

However, the significance of plants as mementos, living beings with their own sets of desires and needs, complicates this idea of an inert yet evocative souvenir. It is one thing for a fragment of pottery or a gemstone to serve as a component of a crafted object, or even a shell, which is at one time home to an organic subject but not itself alive in the typical sense. Does this integration of plant matter and paper (which is, after all, itself derived from plant matter such as hemp or cotton) render the mosaic a more powerfully evocative craft? Does it signify in some ways that this particular type of craft, occupying the borderlands between art and science, has more fully reached its potential as a fusion between the two fields? Of course, by this time in the eighteenth century, women and men had been pressing dried flower and plant pieces into paper volumes for generations (Kew's impressive collection of preserved period herbaria is only a fraction of what was produced), so what makes Delany's mosaics convincingly different? In answering this question, I return again to her correspondence. She writes in October 1774 to her niece Mary Port, "You are saved this time from one of my enormities, by my being in haste to finish a flower for my hortus siccus."[100] While Llanover's footnote remarks, "Unless Mrs. Delany also made a collection of dried plants she must have here again alluded to her great and wonderful work of *the Flora*," it is worth pausing to note that Delany seems to have conceived of her own mosaics as at least, in some respect, a "hortus siccus."[101] In fact, five years later, in a reflective record of her mosaics, she notes, "I at first only meant [them] as an imitation of an hortus siccus," implying that her appreciation and conceptualization of them changed over time.[102]

A collection of dried plant specimens has, at least at base level, a more overtly scientific function than a crafted paper collage. Herbaria of all types have been used to understand changes in plant development, expansion or compression of growing season and regions, and shifting plant DNA. Linnaeus's own herbarium (housed today at the Linnean Society in London) contains over fourteen thousand distinct specimens, including more than four thousand type specimens, and has been used for centuries as a record of eighteenth-century plant life in Europe, the Americas, and Asia.[103] A hortus siccus, however, had a very specialized meaning in the eighteenth century that inhabited the space between personal collecting or craftwork and professional botanical labor. A hortus siccus (literally, "dried garden"), while commonly undertaken by professionals like Linnaeus, evokes strikingly intimate registers. Delany was both building her specimen collection and growing her own personal garden. Further, unlike a more typical dried plant collection, Delany crucially amalgamated her plant specimens into crafted objects of more personal significance, with dimensions beyond the botanical. In naming her project, at least initially, a dried garden, Delany is evoking the vast emotional potential inherent in plants—as individuals for whom she develops a fondness, as exotic, fascinating creatures from abroad, or as one of many inhabitants in her beloved gardens at Bulstrode and elsewhere.

The widespread approbation of Delany's mosaics is a testament to their critical significance in a climate inundated with botanical textbooks, lectures, and works of art. Amid a cacophony of botanical noise, Delany's productions stood out as accomplishing something significant that few of her contemporaries were able to accomplish: a truly original interrogation of the notion of botanical representation. Despite the fact that she adopted few of the standard conventions of botanical artwork and, in many cases, deviated from taxonomic practices, her mosaics were still considered scientifically adept by some of the foremost botanical experts of her generation. Yet, at the same time, Gilpin and Reynolds clearly evaluated them on an artistic scale—a testament to the fluidity of the mosaics' significance and their cultural cachet at a moment when women's botanical work was frequently relegated to strictly domestic spheres. Existing as both domestic objects and public subjects, the mosaics reflect Delany's enormous influence among the eighteenth-century scientific and artistic elite.

The Poetry of the Plants

In closing, I turn to Mary Delany's other, rarely acknowledged botanical productions—her poetry. Such poetry, as limited in scope as it was with respect to her career as a whole, provides an important link between her scientific and literary interests. Written over the course of many years and often referring to her mosaics, these poems also bridge the distance between multiple vectors of her creative pursuits. One of her more notable poems, simply entitled "Plants," was handwritten on a sheet of paper tucked into the first folio of her *Flora Delanica*. This short poem encapsulates not only her emotional connection to her mosaics, but the linkages between her craftwork and her personal relationship with nature. The poem reads in full:

> Hail to the happy hour! when fancy led
> My pensive mind this flow'ry path to tread
> And gave me emulation to presume
> With timid art, to trace fair nature's bloom:
> To view with awe the great Creative Power,
> That shines confess'd, in minutest flower;
> With wonder to pursue the Glorious line,
> And gratefully adore, The Hand Divine![104]

This effusive verse on the subject of her famed paper mosaics, written as prefatory content for a collected volume of the collages, highlights a number of recurring themes around which women naturalists organized their practice.[105] An expression of intellectual satisfaction gained from engaging in scientific pursuit, a popular refrain among women authors and creators, begins the entire poem ("Hail to the happy hour!"). In the next few lines, scientific activity also serves as an emotional balm, ameliorating what might otherwise be a "pensive" state of mind. And, finally, punctuating the entire verse is a common refrain in expressions of women's scientific inquiry: the linkages between botanical contemplation and religious contemplation ("To view with awe the great Creative Power, / That shines confess'd, in minutest flower"). For Delany, scientific activity thereby fulfills multiple needs and accomplishes several goals, rendering botanical study (and, here, scientific creativity) a beneficial path for a woman to wander down.

Read in conjunction with her mosaics, many of them representations of a "minutest flower," this poem also evokes her intimate relationships with plants. As both native and exotic specimens, these specimens support her "timid" efforts to evoke "fair nature's bloom" throughout the years spent in the production of her floral mosaics. The expansion and compression of scale in this poem seems key to Delany's articulation of her mosaic process. Going from "nature's bloom" to "great Creative Power" and then back to "minutest flower," the verse inspires the reader to marvel at the mysteries of the natural world just as its dares her to attempt to visualize them for herself. Readers are called to see both small plants and an unfathomably large deity as part of the same network that forms her paper mosaics and other naturalist craftwork.

Likewise, the "flow'ry path" is not only a metaphorically lovely trajectory but a literal set of motions through garden walkways and estate flower beds. As Delany's craftwork took her across England and Ireland, in and out of many professional and amateur gardens, and virtually around the world, actual physical embodiment and artistic representation become again collapsed. The mosaics as part-plant, part-paper, are both of England and not; their materials sourced often from abroad yet their construction accomplished in domestic spaces, these particular crafts not only "[emulate] . . . fair nature's bloom," they also take on a bloom of their own. As plants that were both the stimulus for and the building blocks of the *Flora Delanica*, these blossoms are indeed worthy of this laudatory verse.

One of Delany's most iconic mosaic creations, the Portlandia grandiflora, was important enough to Delany herself to merit its own brief poem, contained again in the correspondence provided by Llanover. Like the poem examined above, this verse is designed to honor both her mosaic project and something greater than the melding of paper and plant; in this case, her best friend the Duchess of Portland. It reads in its entirety:

Fair flower! that bears the honoured name
Of HER whose fair and spotless fame
  Thy purity displays.
Emblem of Friendship's sacred tie,
Thy form is graced with dignity
  Superior to all praise.[106]

As with her articulation of the relationship between nature and art, here Delany incorporates friendship into this dynamic through language evocative of deep emotion and even piety. The "fair flower" is centralized at the beginning of the verse, but shortly thereafter assumes a secondary role as Delany pursues an encomium on her beloved friend. The Duchess's "fair and spotless fame" evokes the purity of the Portlandia, just as the blossom's lines and curves convey the impression of dignity and superiority. Friendship between women is not only personally significant to Delany but achieves a broader cultural and spiritual consequence in this poem: "Friendship's sacred tie" sanctifies female intimacies in keeping with a long tradition of female friendship poetry. A writer like Katherine Phillips, for example, writing decades before Delany's birth, had achieved the reputation of a "matchless Orinda" for her own "fair and spotless fame" as she, too, articulated deep female intimacies.

What is key in this particular poem is the way those deep female intimacies become intertwined with botanical knowledge and networks. The mosaic is dated August 9, 1782, making it one of the last mosaics she ever completed; it is certainly among the best known (fig. 7). The Portlandia also represents a significant convergence of aristocratic influence and privilege: the plant itself was transplanted from Jamaica to be grown at the Royal Gardens at Kew, a specimen was specifically sent to Delany by Queen Charlotte, and the mosaic was completed at its namesake's home in Bulstrode. In this way, the poem's movement from flower to woman, from plant body to female body, mimics the movement of both women and plants in broader networks both global and domestic. The name of "Portlandia" was also an evident testament to the influence wielded by the Duchess in scientific circles and, by extension, Delany herself. As one of the first people in Britain to depict the Portlandia, Delany is no less important a personage in the history of this exotic specimen than its namesake. Further, the decision to list only the Latin name of the plant and not also its vulgar name, bell flower, suggests an important conviction of the significance of female presence. The plant was even valuable enough to Delany for the mosaic to be accompanied by an explanatory

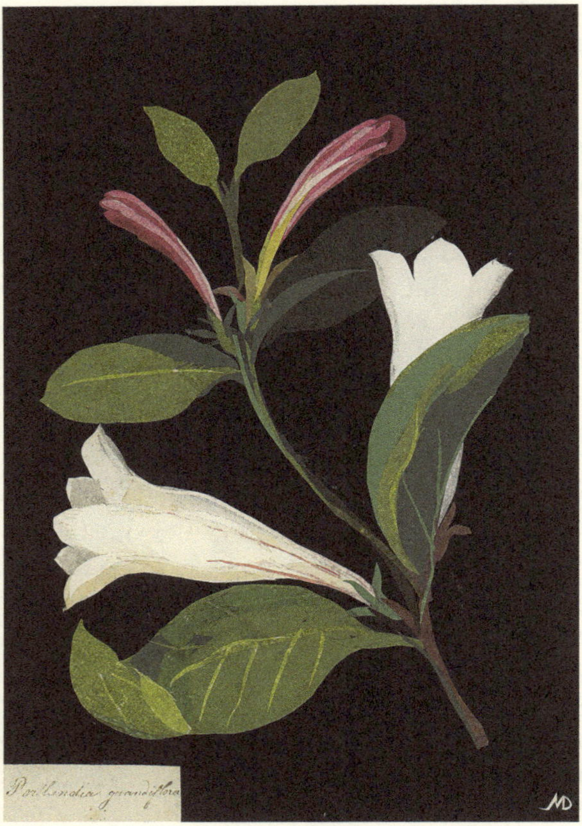

FIGURE 7. "*Portlandia grandiflora*," Mary Delany, 1782. (© The Trustees of the British Museum)

sheet that cited "Brown's *History of Jamaica*" as evidence for its origins—a contextualization rarely provided for the mosaics. This convergence of materials and information around the Portlandia highlights Delany's active scientific and creative endeavors across multiple vectors. The accompanying poem thereby bolsters not only the relationship between the two women but their roles as naturalists as well.

The mosaic itself is one of her simpler compositions, with fewer fine pieces of paper and larger blocks of the same color than other mosaics completed around the same time. Simultaneously, however, the composition mimics the sentiments expressed in the poem: the blossoms, lovely and fair, are turned away from the viewer in a posture that conjures dignity. The mosaic also evokes a kind of botanical purity through Delany's decision not to depict the reproductive components of the plant, as one would see with the blossoms facing head-on. Because Delany did not shy from depicting such fructifying components in other mosaics (the *Passiflora laurifolia* being one of the most remarkable in this respect), the decision to abstain from doing so in this one seems purposeful. In fact, an illustration of the same plant completed by James Sowerby for James Edward Smith's *Icones pictae plantarum rariorum descriptionibus et observationibus illustratae* (1790–93) features markedly recognizable stamens poking out of the petals (fig. 8). Sowerby's image, while more detailed by virtue of being an engraved plate from a very expensive book, is also muted in color by comparison to Delany's Portlandia, making the latter's image both more "spotless" and more "fair." By departing from conventions of botanical illustration—even conventions she had adhered to previously—Delany created a remarkable depiction of the Portlandia that was deserving of her admiring verse.

On the other end of the emotional spectrum, Delany also turned to poetry when her mosaic project could no longer be continued due to her failing eyesight. This poem, written lamenting the end of the *Flora Delanica* in 1782, summarizes the various intellectual and emotional rewards the construction of the mosaics brought for Delany. The creation of the mosaics is here represented as attractive, enticing, and holistically fulfilling, with emotional responses at the highest registers:

> The time is come! I can no more
> The vegetable world explore;
> No more with rapture cull each flower
> That paints the mead or twines the bower;
> No more with admiration see
> Its beauteous form and symmetry!
> No more attempt with hope elate
> Its lovely hues to imitate!
> Farewell! to all those friendly powers
> That *blest* my solitary hours.[107]

Here, her floral specimens evoke a variety of emotions that can ambiguously occupy a space between sensuality and platonic friendship. As they have been on occasion in her correspondence, the mosaics are explicitly feminized, described as "beauteous," "lovely," and symmetrical. The poem also depicts the mosaics' dual function as things and things to do: the mosaics, as objects, are attractive, and, as an activity, elating and rapturous. While other poetry of the period might imply these "raptures" to be sensual ones, as I have noted, her use of the word "rapture" regularly

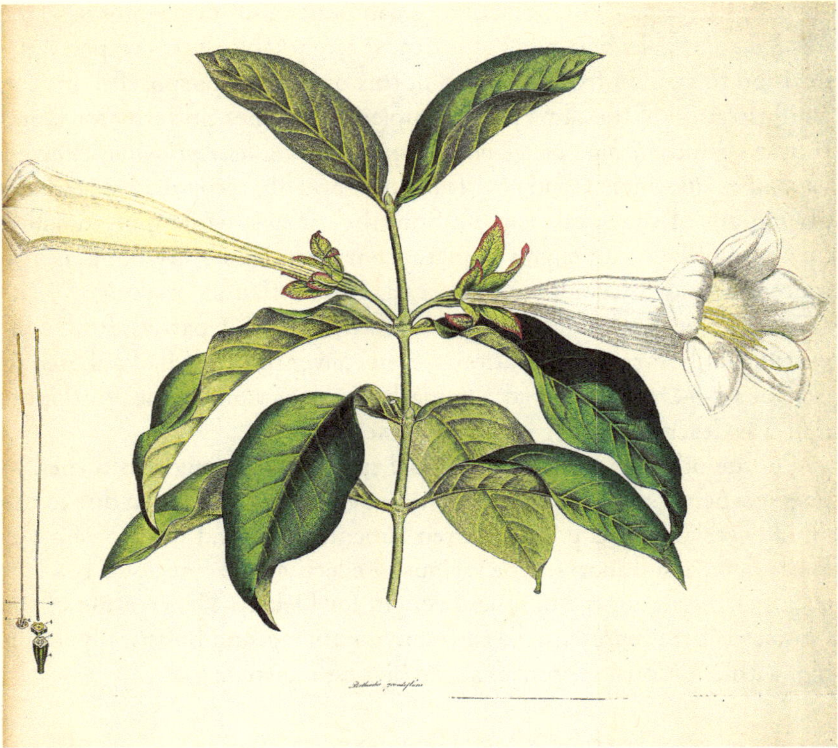

FIGURE 8. *"Portlandia grandiflora,"* James Sowerby, 1790. (Image from the Biodiversity Heritage Library; contributed by Missouri Botanical Garden, Peter H. Raven Library)

conveys spiritual rather than sexual overtones. However, there is nevertheless something more than exuberant piety at play in these verses: the pursuit of botanical study and its accompanying craftwork has been for Delany a source of pure and unadulterated joy.

I also want to highlight Delany's conceptualization of the way she processed the flowers into mosaics: the precise phrase used is "cull each flower / That paints the mead or twines the bower." While this phrasing evokes the unremarkable action of picking a flower, it also implies a kind of selectivity. Flowers are "culled," presumably from a larger population, for various qualities that make them worthy of a space in her "invaluable folios." Rather than picking every honeysuckle or morning glory that "twines the bower" and using them as botanical subjects, Delany carefully selects her blossoms from the "vegetable world" for rarity, beauty, symmetry, or other noteworthy qualities. Most important, those noteworthy qualities may be judged entirely at her discretion, rather than in adherence to a more technical or professional botanical assessment of their value. For example, the total collection of mosaics features over thirty-five different kinds of geraniums (nine of which featured actual plant samples). As recently acclimatized plants from temperate parts of Asia and the Levant, geraniums were horticultural darlings but were often critiqued by professional botanists due to the frequent interference of horticultural techniques in their growth and reproduction. Delany, conversely, found these exotic plants engaging enough to incorporate actual components of their organic bodies into the mosaics. Geraniums were only one among many repeated species in her collection, too: others included ranunculus, anemone, and hyacinth—all "florist's flowers" that few botanists approved of.[108] In this respect, Delany's scientific craftwork departed from professional conventions in pursuit of her own botanical inspiration. As much as historians struggle to find an organizing conceit for the *Flora Delanica,* the best characterization may be to say that she pursued artistic representations of plants that interested her.

The poem concludes on a different note, a sad "farewell!" that leads into Delany's preferred mode of reflecting on the natural world—through theology. The poem continues:

Alas! farewell! but shall I mourn?
As one who is of hope forlorn?
Ah no! my mind with rapture feels
The promise which thy Word reveals.
Come Holy Spirit, on thy wing
Thy sacred consolation bring.
. . . .
Teach me, *submissive to resign*
When summoned by Thy Will Divine.[109]

The "sacred consolation" in bending to "Thy Will Divine" seems to no longer be about Delany's relationship with plants but instead her relationship with a broader conceptualization of nature. More specifically, it emphasizes her relationship with the creator of nature—the holy being who had brought her to raptures over her shells as a young woman, serenity among the Duchess's zoo later in life, and now to peace, with the end of her own story. While she may mourn the loss of the mosaic artwork and, at this point, many of her other forms of craftwork as well, she found new inspiration by looking to her faith. While Delany would live another six years after penning this verse, as well as continue to actively correspond with her friends and family members, she never completed any more mosaics. Drawing to a close as it had begun, Delany's natural history craftwork remained rooted in the appreciation of the creator of nature. As she wrote nearly two decades earlier, "Surely an application to natural beauties must enlarge the mind? Can we view the wonderful texture of every leaf and flower, the dazzling and varied plumage of birds . . . &c. &c., and their infinite variety, without saying, '*Wonderful and marvellous art thou in all thy works!*'"[110] The pursuit of natural history, for Delany, was always about something more than a fascination with botany or an interest in collecting shells: it incorporated and expanded those passions as a part of her natural theology.

Delany was only one of many individuals who were inspired by the study of botany to create new, innovative ways of engaging with the field that ultimately transcended systematized botanical knowledge: from Anna Seward's poetry to Thornton's exquisite *Temple of Flora,* creative interpretations of natural history formed a lively discourse of their own in the period. This trend suggests that Mary Delany was less of an outlier and more of an outstanding representative of the creative possibilities of botanical knowledge organization in the period that were emerging from the amateur scientific culture, many of which were projects by women. Like Rowden and Hoare, Charlotte Smith also wrote poetry with botanical elements: "Beachy Head" is filled with taxonomic footnotes and innovative explorations of the relationship between human and plant life (as is "Flora," the subject of the upcoming coda to this book). Delany herself also inspired a number of female successors in her mosaic artform, including a Mrs. North and her companion Miss Hamilton.[111] And so new opportunities for rewarding intellectual labor blossom at the places where science and art converge—in craftwork, in poetry, and in the floral mosaic.

Mary Delany's oeuvre touches upon nearly every one of the variable scientific venues open for female participation in the late eighteenth century; from botany and conchology to craftwork and artwork, her interactions with the natural world were consistently influenced by her distinctive approach to scientific study. Looking to nature as a source of inspiration and mental stimulation, she used each of her productions to articulate her unique vision of the relationship between women, nature, and natural history. Through her collections of plants, shells, and minerals, Delany cultivated relationships with women all across Britain; through her craftwork she further nurtured and developed those relationships. As she utilized these items of scientific craftwork in female networks of exchange, she participated in a scientific discourse that distinguished itself through creativity, intimacy, and ingenuity. Finally, in inventing the floral mosaic, Delany created new possibilities for what botanical artwork could accomplish. By working within scientific conventions but subtly reimagining what those conventions might look like as expressed by a woman, she offers up an alternative model of scientific practice that impressed and inspired her contemporaries and her successors alike. The women who embraced her legacy of creative scientific praxis can be found across multiple literary and artistic fields; not too long after Delany envisioned alternate modes of relating to vegetality through innovative craftwork, Charlotte Smith ambitiously reworked the poetic possibilities of woman-plant relationships.

The upcoming coda, an analysis of Smith's poetic articulation of vegetal proliferation, comprises the final section of *Botanical Entanglements*. Building upon the intimate explorations of the relationship between women and plants undertaken by Elizabeth Blackwell and Henrietta Maria Moriarty while returning to a focus on literature, this segment also links to Delany's investment in poetic vegetal expression. Through a sustained close reading of a single poem—"Flora"—I highlight the author's sensitivity to plants as beings with their own needs, desires, and ambitions. In chapter 2 I discussed Maria Jacson's musings upon the possibility of plants exhibiting human-like personalities and Priscilla Wakefield's explication of the analogist model of plant/human corporeality; the coda showcases Smith's astute combination of both concepts in a compelling poetic form. I also offer an interpretation of Smith's poetic botanical praxis that illuminates her subversive engagement with both naturalist and literary forms.

Coda

> I wish you to learn the names of all the trees, shrubs, flowers, and vegetables, which are every day before your eyes, and . . . to have delight in being acquainted with the indigenous or native plants of the country you live in.
>
> —Charlotte Smith, *Minor Morals*

In this concluding section, I look to a single poem by Charlotte Smith that elucidates the author's vision of the relationship between botany and poetry as well as her distinctive and often complicated deployment of taxonomic data in the service of poetic authority. Published originally as part of *Conversations Introducing Poetry* and released again with *Beachy Head and Other Works,* the short pastoral poem "Flora" is usually overlooked in criticism of Smith's botanic verse in favor of the lengthier title track. Critics like Melissa Bailes, Kandi Tayebi, Donelle R. Ruwe, and Theresa Kelley have all argued in favor of "Beachy Head" as indicative of the author's deep investment in the convergence between poetic and scientific authority, specifically with reference to her deployment of extensive naturalist endnotes that correspond to plants and animals mentioned within the lines of the verse itself. A closer examination of "Flora," however, undermines the assertion that endnotes served a clarifying or bolstering function in Smith's poetry: while botanic notes do indeed underpin her references to various biota in the poem, I argue that they fail to do so in a coherent and consistent way, thereby evoking the true, riotous essence of verdant plant life.

Although she was hardly the only Romantic poet writing about plants, Smith's engagement with vegetality was rooted in her enduring passion for the technical details of scientific subjects. Her naturalist writing career was remarkably diverse, featuring a variety of texts, tactics, and ideas drawn from botany, natural history, and ornithology. For one, her 1796 novel *Marchmont* depicted the young protagonist Althea Dacres as an amateur flower painter, acknowledging in fiction the value that creative engagement with the natural world held for women. She also wrote two educational treatises for children of both genders, *Rural Walks* (1795) and its continuation *Rambles Farther* (1796), which focused on the lessons that children might extrapolate from small bits of information about the natural world, such as the reproduction of plants or the lifecycle of insects (*Minor Morals* [1798] is typically grouped with these texts as well). Later, she also published *Conversations Introducing Poetry: Chiefly on Subjects of Natural History* (1804), which concretizes her theory of the relationship between poetry and science that would later emerge as a central principle behind her most famous poem, the ostensibly unfinished "Beachy Head" (1807). Finally, with a specific focus on ornithology and its cultural significance as a field of study, *The Natural History of Birds* (1807) was released by her publisher only after her death.

I have selected the under-studied poem "Flora" for the subject of this coda as a compact example of Smith's evolving ideas about scientific documentation and vegetal assessment. Tayebi has proposed that "all of the footnotes [in 'Flora'] lead the reader toward the idea that there is an underlying order and classification to the overflowing abundance of natural phenomena in the world," which is consonant with the typical interpretation of taxonomy as essentially a clarifying process.[1] As Kelley explains, "During the Romantic era . . . 'new' or exotic flora and continued scrutiny of newly discovered aspects of known flora put a continuous pressure on the desire to classify, reclassify, and name species and systematic arrangements to which species were said to belong."[2] However, as the eighteenth century wore on, it became increasingly clear to English and European naturalists that Linnaean taxonomy was, to use Maria Jacson's term, "generally arbitrary" at times, with its singular focus on the reproductive organs of the plant at the cost of excluding all other morphological features of the organism. Furthermore, with its explicit focus on flowers, a population of organisms to which humans have attributed expansive and eclectic cultural meaning, Linnaean taxonomy ironically works to defeat its own sterilizing ends. As Isabel Kranz reminds us, "By assigning flowers the role of ordering the plant kingdom, thus making them the key indicator for systemic and structured knowledge . . . an esoteric undercurrent emerges in which flowers convey a surplus of meaning that cannot be contained."[3] Thus, while botanists and theorists throughout the Romantic period and into the Victorian era would begin to contest

Linnaean taxonomy and replace it with the idea of "Natural Orders," it was also the case that the very idea of taxonomy as a stabilizing and orderly force was not always plausible.[4]

Michael Marder has proposed that the application of taxonomic naming strategies to plants is oppressive and destructive, rather than illuminating. He elaborates: "These [taxonomic] names are meant to capture the essence of the plant by assigning to it an exact phrase in a dead, albeit highly differentiated, system that swallows up the [plant's] singularity and uniqueness."[5] That is, while the taxonomic system achieves its delimited goal of imposing a kind of order on the richness of the natural world, it ultimately renders each individual plant "nothing in itself outside the intricate net of classifications wherein it is caught up"; plants are no longer singular organisms with their own significance and place in the environment, but are simply valuable for their adherence to a type specimen.[6] This attempted reduction of what is essentially the irreducible nature of plant life (no two plants are the same, from morphological, chemical, and genetic standpoints) must thereby ultimately fail in its attempt at imposing a coherent scheme on an incoherent ecosystem.

Late eighteenth-century botanists understood Linnaean taxonomy to be flawed—as, to be fair, Linnaeus himself recognized—albeit while continuing to insist upon the viability of taxonomy as a metric for understanding nature. Critics rightly perceived Linnaeus's singular focus on sexual organs as an artificial system of botanical categorization and, following the death of Linnean Society President James E. Smith in 1828, most professional English botanists began to prefer the naturalist system of the French botanist Antoine Laurent de Jussieu.[7] This botanical sea change contributed to the diminished popularity of Linnaean taxonomy as a scientific system in the nineteenth century, which unfortunately also lead to an increasingly professionalized scientific climate that no longer embraced amateur participation (and therefore narrowed the opportunities women had for exhibiting scientific authority). While the boundaries of this particular study do not extend beyond Smith, a brief review of women's botanical writing in the nineteenth century indicates that a significant shift away from Linnaean taxonomy was accompanied by a reduction in the use of precise scientific terms and concepts, which in turn inspired most published naturalist texts by women to deflect nearly any kind of scientific expertise. This gives rise to gardening manuals with little to no botanical information (including several by Jane Loudon), didactic works that purposefully mocked women with scientific pretensions, and illustrated guides to the language of flowers like the ones produced by Anna Christian Burke in the second half of the century. Such a trend also essentially brought an end to the nuanced inclusion of scientific information in well-regarded published poetry: for example, Bailes has shown how Felicia Hemans wrote with a purposefully nontechnical specificity in her botanical poetry in order to avoid accusations of intellectual impropriety and to portray Linnaean systematics as antiquated.[8] While there is little evidence to suggest that any naturalist of the period actively rejected the entire premise of taxonomy, Charlotte Smith's subversive deployment of prevailing taxonomic schemas highlights the potential for those outside of conventional structures of authority to expose the weaknesses of scientific orthodoxy.

Just as the arbitrary nature of Linnaean taxonomy proved unsustainable at the end of the eighteenth century, modern versions of taxonomy encounter an increasing number of specimens in new environments (new only to Western scientists, usually) that strain and stress current taxonomic models—hence the seemingly endless proliferation of taxa in modern botanical systems. This unresolved and unresolvable tension between taxonomy's stated aims and its inevitable failure is embodied in Smith's "Flora." The poem itself begins with a typical pastoral reflection, proposing the fantasy of "Flora" as an alternative to the trials and sorrows of a difficult life. The "enchanting goddess of the flowery tribe"—the eponymous Flora—arrives in an energetic vision of abundant nature that has the power to "soothe the wearied pilgrim's eyes" long after the vision has faded.[9] The poem continues with a litany of flowers, trees, insects, and even aquatic plants, envisioning a beautiful, multivalent garden that represents the best of nature's bounty. The world of the poem is also a world in which everything has its place: the goddess Flora presides over all, and various sylphs, naiads, and nereids care for the natural world at her direction. However, this pastoral fantasy is unsustainable. Not merely, as Tayebi has argued, because of the presence of militant sylphs fighting off invasive insects ("For conquest arm'd these pigmy warriors wield / The thorny lance, and spread the hollow shield / Of lichen tough"), but because of Smith's purposeful mixing of Latinate/Linnaean terminology and common names for flora and fauna (61–63).[10] Although initially a poetic decision, the choice to mix terminology and registers is also a scientific one.

Judith Pascoe suggests that this tactic reflects "intellectual as well as artistic assertion," interpreting Smith's deployment of scientific specificity as a particular kind of feminine authority.[11] While it is true that Smith utilized the genre of the endnote in her poetry in part to substantiate her presentation of scientific information (and her right as a woman to serve as an authority on such knowledge), it seems equally significant that the authority she conveys is far from absolute. Notes exist for some terms and not for others; they provide dramatically varying levels of detail, ranging from the Linnaean name only to personal asides on the plant or animal in question; they occasionally document sources for the information, but not consistently. In this, Smith's deployment of endnotes is similar to James Grainger's use of scientific notes in his lengthy West Indian georgic, *The Sugar-Cane* (1764), in which the poetic expression of exotic indigenous plant life is supplemented by additional scientific information like Latinate terms, growth patterns, geographic distribution, and cultural uses. While Smith avoids the exhaustive type of note found in Grainger's explication in book 1 of "the spiry Cane" (a note that spans several pages of a global history of cultivation ranging from the earliest mentions of the plant in Greek and Hebrew texts to its cultivation in the New World in the sixteenth and seventeenth centuries), it is useful to acknowledge that her utilization of scientific endnotes has a respected history in English literary culture.[12]

Admittedly, a poem is neither required nor expected to maintain a level of scientific rigor that one might expect from a textbook, for example, but Smith's usage of the notes seems at best partial. A passage in the middle of the poem is illuminating on this point, as she nuances the setting of her vernal fantasy with a pairing of different styles of classification:

> Each forest thicket feels the balmy air,
> And plants that love the shade are blowing there.
> Rude rocks with Filices and Bryums smile,
> And wastes are gay with Thyme and Chamomile. (135–38)

Filices (ferns) and Bryums (a specific genus of mosses) are Latinate names without being overly specific, but they do require a certain familiarity with botany and natural history in a way that many of the other plant references do not. And, unlike most of her other Latinate references, she does not include any of her own editorial notes (although modern editors usually provide them). Without endnotes to these two terms, which are found well into the poem following multiple earlier endnotes, the function of the notes as a textual apparatus seems confused. Most readers would recognize thyme and chamomile, but to pair such common terms for common plants with uncommon terms for other common plants is to create an uneven ground upon which to evaluate the species contained within the poem. What most critics have identified as Smith's intention to provide contextual information for her readers while underscoring her own scientific authority is therefore not fully substantiated by the deployment of taxonomy in "Flora."

This habit of grouping different kinds of names is found throughout the poem. For example, the earliest of the flora species mentioned is "Saxifrage," elaborated in Smith's note as "Saxifraga hyniodes—Moss Saxifrage, commonly called Lady's Cushion" (33). The confluence of terms and ideas for a single plant—the Linnaean term, the vulgar term, and a local or regional term—evokes for the reader not an idea of specificity, but of verdant abundance. Three names for one plant, while certainly edifying, is also overwhelming. Like many botanical texts of the period, from illustrated herbals like Elizabeth Blackwell's, which itemized up to a dozen names for a single plant in one entry, to the perpetual doubling of vulgar/Latinate terms in Jacson's *Botanical Dialogues*, "Flora" actively participates in this multiplicity, veering through the factual (it is helpful to know multiple names for a plant) into the chaotic (this plant is simultaneously all of these names and none of these names).

The doubling and tripling continue throughout the next several lines, as Smith introduces first "tufted rush" and then "convolvuli" in text, with the Linnaean terms in notes, "Eriophorum angustifolium" and "Convolvulus arvensis" (37, 41). While it is true that "convolvulus" was a familiar enough Linnaean term to become the more dominant name for the plant, later lines undermine the theory that Smith was simply using common names for plants whose Latinate names were less well known. "Scandix" and "Anthoxanthum" were more likely to have been known by their regional or vulgar appellations, "Venus' comb" and "Vernal Meadow Grass," which Smith provides in her endnotes (45, 48). The image of the Scandix, furthermore, blossoms into four different names for one plant: Scandix, Scandix pecten, Venus' comb, and Shepherd's needle. Depicted as the plant containing Flora's flowing hair, the choice of "Venus' comb" as an ornament seems especially apt, but why not use the vulgar name in the verse? In her decision to relegate the more ostensibly poetic term to the note, Smith's poetic philosophy seemingly ignores the advice given by John Aikin in *An Essay on the Application of Natural History to Poetry* (1777), which encouraged poets to choose the appellation for plant and animal life with the most sonic and artistic potential.[13] Rather than a neglect of style, however, her use of multiple names for her plant specimens instead indicates her embrace of the inherent complexity of plant life;

for Smith, fully understanding the image of "wandering tresses" bound with flowers relies upon the multiple layering of taxa and terms. Consequently, comprehension of her poetic imagery is based upon a shared body of scientific knowledge as well as an appreciation for the multiplicity that such layering ensures (43).

Lines 63 through 84 abound with plants whose names are layered like petals and leaves. Swerving between "foxglove"/"Digitalis purpurea" and "Leontodon"/"Dent-de-lion," Smith switches the hierarchy between common name and Latinate term seemingly without logic (65, 75). "Dandelion" is obviously the more common term, as is "foxglove," but only the latter is privileged in the text, where the former is oddly obscured through two additional layers of naming, as Smith proceeds through "Leontodon" to "Leontodon officinalis" and only then to "Common Dent-de-lion." Even more significantly, "dandelion" was the preferred spelling in the eighteenth century (Blackwell used it in her herbal more than seventy years prior), adding a fourth, implied layer for Smith's readers. She provides the same inconsistent layering with "Arum" and "Burdock," preferring the Latinate term in favor of the vulgar "Cuckoo pint, or Lords and Ladies" in the first and using the vulgar term rather than the Linnaean one in the second (70–71). By providing not one but two regional names for the Arum, Smith's attempt at clarity achieves instead plurality. The plants are both singular (types, identified by specific names and serving specific narrative functions in the poem itself) and multiple (they have too many names, and slip out of their coherence as individual organic objects). It is here where I identify "Flora" as exemplifying what Lara Farina calls the "tendency [of plants] to *mass* in Western perception and thought."[14]

Farina makes the argument that by looking to medieval texts, we notice the ways in which "plant matter worked against linguistic species-marking by clustering into *masses*, forms that override difference and thus resist closure."[15] Not only is this also accurate with respect to eighteenth-century texts, it is especially apt in Smith's "Flora." The poem is filled with plant names, Latinate and vulgar, and the plant bodies behind them. It is also full of fairies whose names are inspired by plant anatomy: "Floscella," "Petella," "Nectarynia," and so on (81–91). Finally, the entire poem is predicated upon the arrival of the goddess of plant life, Flora, who is garbed in plants and may be part plant herself. This is a mass of vegetal life comprising, and then overwhelming, the pastoral boundaries of poetic form. Like the eighteenth-century herbals that Smith surely used as reference in writing "Flora" (alongside the texts of Thomas Martyn, William Withering, and Colin Milne), the poem ends up with a messy pile of names whose ostensible containment in endnotes and asides eventually manages only to reinforce the untenable nature of taxonomy as a whole. If one name for a plant isn't enough (and it never is), then the entire project is unsustainable. Taxonomic naming may serve a putative organizing function, but it cannot truly contain or order plant life as it exists either in nature or in pastoral fantasy.

Smith later even abandons her attempts at taxonomic notation in subsequent sections, including a note to line 101 that simply reads, "The following lines describing well known flowers, notes would be superfluous." Vegetal life considered "well known" includes "Hyacinths," "Crown imperial," "Gentian," "Jonquils," "Violets," "Honeysuckle," "Rose," and several tree species (112–21). But if the function of those taxonomic endnotes is, at least in part, an attempt to educate her readers about the Latinate and Linnaean terms for plants (as well as an attempt to demonstrate her own familiarity with botanical science), why forgo them and then continue to use only common terms for plants? Only a few of these species are referred to by anything approaching their Latinate names, and the remainder are left with singular points of reference which inhabit a space before and outside of taxonomic structures of organization (e.g., the rose was a rose long before Linnaeus and continued to be a rose long after). The status of being "well known" also seems to apply to acclimatized as well as indigenous plants—the crown imperial was native to the Levant, while the gentian was common in Central Europe, and violets grow abundantly in multiple climes. What, precisely, about these specific species makes them unworthy of serving a pedagogical function for Smith's readers? Perhaps, as the saying goes, familiarity breeds contempt—or, possibly, and more likely, these plants have been absorbed into Western botanical knowledge long enough to have lost some of that proliferating abundance that exotic specimens exhibited with gusto.

Yet their very presence in this pastoral fantasy, with mixed registers of native and exotic specimens, Latinate and common names, also pulls them into the overarching sense of vegetal profusion. Smith's refusal to confine the bulk of the plants in "Flora" to a single name is thereby much more than a pedagogical tactic inspired by her *Rural Walks, Rambles Farther,* and other naturalist writings. Instead, it reflects the ultimate inability of her endnotes to encapsulate the true essence of the plants whose images she evokes. This accumulation of different names, different species, different plants—all of it "masses" in ways that resist tidy categorization. In this way, the ostensibly artificial garden grown in a poem like "Flora," a pastoral imagining that pulls together plants from a wide geographical range and a broad span of cultural signification, becomes more like a real garden than we might at first think. The plants of "Flora," like real flora, proliferate, grow, and expand out of the author's control. As Marijke van der Veen reminds us, "Plants have their own agenda . . . [which is] to make copies of themselves," to continually increase in size and in number.[16] Marder refers to this as the "bad infinity" of plant life, or the capacity for endless growth that often makes humans, species with discrete corporeal boundaries (at least in theory), somewhat uncomfortable.[17]

Plants and people, however, have more in common than otherwise, as Smith herself recognizes. While the "people" in the world of "Flora" are actually fairies, they blur the boundaries between the humanoid body and the vegetal body to dramatic effect. The above-mentioned sylphs who guard Flora's vegetal children against intruding insects are armed and garbed with plant parts: they carry swords made of thorns and shields made from lichen and wear foxglove and Scutellaria ("small scull-cap," vulgarly) as armor (63–66). Others bear "clubs" made of arum and "hooks" from burdock and wear "[casques]" of dandelion blossoms (67–71). On one level a quotidian depiction of woodland fairies wreathed in plants and flowers, it also reminds Smith's readers that the complex relationship between the human and vegetal worlds is always a mutually dependent one. Plants provide clothing and defense mechanisms for the sylphs, just as the fairies then use those plants to defend other plants. This imbrication of plant and human/fae is only deepened when Smith names a few of her guardian fairies—Calyxa, Nectarynia, Petalla (87–97). As I have mentioned, these names are feminine adaptations of Latinate names for plant anatomy: in these cases, calyx, nectary, and petal. These humanoid creatures do not just wear plant parts; they *are*, in some sense, plant parts or part plant.

The morphological similarities between humans and plants were of great interest to eighteenth-century naturalists (including women), as discussed in chapter 2. Erasmus Darwin and Priscilla Wakefield, among many others, raised questions about plant biology that equated vegetal and human biology with respect to concepts like respiration and circulation. For example, Wakefield noted that bark served the same function as skin, and that plants were said to "breathe" just like animals.[18] Darwin observed that plants could be said to have a brain because they possessed the capacity for volitional movement and even sleep.[19] The plant body's biological and psychological similarities to the human body thus evoked comparisons that pulled humans closer into the vegetal world—if plants could think, feel, and react, then they seemed to exhibit human qualities. Or, said another way, humans, as the newer organisms on the planet, exhibit plant qualities. While eighteenth-century thinkers usually retained the hierarchical valuation of organic life that placed vegetality at the bottom and humanity at the top (following Aristotle's philosophical ordering of conscious lifeforms in *De Anima*), it is also the case that period hierarchies placed women at the bottom and men at the top. Yoked together as inferiors in the grand scheme of the universe, women were able to cultivate strategic alliances with plants for mutual purposes. Of course, as Smith has shown us, humans and plants are already mutually entangled. Plants gain more water, better sunlight, greater reproductive opportunities, and expanded growth thanks to gardeners and farmers; people gain food, clothing, shelter, oxygen, medicine, and other material resources through the cultivation and harvesting of various plants. Women are part of this dynamic, but they are also part of what I have been arguing is a distinct dynamic in which their unique relationality to nature allows them a kind of special opportunity to relate to plant (and animal) life in different ways.

As I have shown throughout *Botanical Entanglements*, women could find edification and emotional fulfillment by tending to small insects, grow their own gardens to exchange seeds and specimens with other women, produce texts through which fellow women could intimately encounter and learn about botany and natural history, or create new ways of engaging with the organic world through naturalist artistry and craftwork. By underscoring the significance of such connections, we see how Smith's role as an educator and a naturalist is also underpinned by her biocultural position as a woman. I am not arguing that Smith's taxonomic project in "Flora" failed because, as a woman, she was not the one making or naming taxa. I do think that it failed, however, because she imperfectly inhabited the frame of mind—botanic, poetic—necessary for taxonomic labor to achieve its purpose. Proliferation of names and plants is, in fact, the opposite of what taxonomy attempts to accomplish. By her very understanding of the need to acknowledge the multiple identities of plants, Smith renders her own efforts admirable but futile. Through a reading of "Flora" that acknowledges the inexorable multiplicity of vegetality, we better understand how Smith's poetic vision is about much more than quirky endnotes. Rather, it is about a celebration of plant life, of its abundance, of its mass, and of its ultimate unknowability. This is perhaps the greatest lesson that Charlotte Smith has to teach her readers: that vegetality is ultimately beyond human ken.

Highlighting the fact that the authors discussed in this project—Eliza Haywood, Charlotte Lennox, Priscilla Wakefield, Maria Jacson, Maria Edgeworth, Elizabeth Blackwell, Henrietta Maria Moriarty, Mary Delany, and, now, Charlotte Smith—all struggled to understand vegetality was among the central aims of this book. Typical ways of relating to nature available to men were often unavailable to women, so, in order to understand and relate to the natural world, they strategically manipulated the distinctive convergences between femininity, domesticity, art, literature, and science. By employing their unique perspectives on the plants, animals, and insects that inhabited the gardens of England, these women effectively resisted the detrimental linkages between themselves and the natural world and, instead, grew their own relationships. Through a diversity of strategies, formats, texts, and topics, these women all cultivated a personal connection with nature: not a theoretically perfect connection, full of complete knowledge or dominion, but a deeply intimate one, bursting with promise and potential. If the goal of our modern reconceptualizations of the affiliation between humans and the natural world is to reinvigorate what it might mean to think of vegetality and nature as deeply imbricated with humanity and human life, rather than ancillary to it, then we must acknowledge that these female authors and artists began that project a long time ago.

As literary critics, historians of science, and parties interested in thinking about the future of human-plant entanglements, recognizing the work of our predecessors is an important first step. *Botanical Entanglements* has attempted to not only reshape the archive of women's scientific labor in the eighteenth century, but to offer inspiration for the path forward. There is still, of course, plenty to be done in terms of rethinking the dynamics between humanity and vegetality. However, whether we ponder the broad implications of plant intelligence, the role of humans in destroying an increasingly fragile environment, or the urgent need for a radical change in global environmental policies, we must always remember the relationship as one of mutual indebtedness. The "environment" is not an inert space for our pleasure and utility or a convenient backdrop to human activity, but a complex ecosystem of plants, animals, microbes, soil, water, and atmosphere fully possessed of its own dynamics and desires. Further, we need plants in a way that, ultimately, most plants don't need us. If we remember that there is much to be learned from nature, and that nature can hold both inspiration and joy for curious individuals—as the women of this book understood intimately—perhaps we can also remember our own responsibility to the world in which we live. After all, in spite of our most ambitious speculative dreams, as far as we know, humans only get one try at this.

NOTES

Introduction

1. There were a few early groups that met more casually under the auspices of studying and discussing botany, including the Temple Coffee-house Botanical Club (founded in 1689). Additional professional societies with an exclusive focus on botany were formed later in the century, given the disparity in perceived degrees of rigor and significance between, for example, chemistry and botany. This was directly correlated to the extensive record of amateur participation in botanical study that encouraged women and laypersons to contribute. Botanical societies of note included the tiny Lichfield Botanical Society and the more expansive Linnean Society of London, the latter of which is still flourishing today.
2. To provide only examples of their earliest work on the subject: Merchant, *Death of Nature;* Haraway, *Primate Visions* and *Simians, Cyborgs, and Women;* Harding, *Science Question in Feminism;* Keller, *Reflections on Gender and Science* and *Secrets of Life/Secrets of Death.*
3. Shteir, *Cultivating Women;* Schiebinger, *Mind Has No Sex?* and *Nature's Body;* Gates, *Kindred Nature;* Phillips, *Scientific Lady;* and Benjamin, *Question of Identity.*
4. For a sustained study of female participation in (and reaction to) botanical discourses, see George, *Botany, Sexuality, and Women's Writing.* George's argument, however, relies upon a binaristic approach to the history of science which more recent work has softened.
5. See Tobin, *Duchess's Shells.*
6. Studies of women like Maria Riddell, Maria Graham, Jane Colden, and Mary Anning, for example, have usefully contributed to a fuller portrait of naturalist productions in the period.
7. See Leong, *Recipes and Everyday Knowledge;* Leong and Rankin, *Secrets and Knowledge;* Smith, Meyers, and Cook, eds., *Ways of Making and Knowing;* Smith, *Entangled Itineraries;* and Wall, *Recipes for Thought.*
8. Bailes, *Questioning Nature;* and Gevirtz, *Women, the Novel, and Natural Philosophy.*
9. Significant recent volumes on botanical artistry (with varying degrees of focus on women's work in the genre) include Kramer, *Women of Flowers;* Keeler, *Gardens in Perpetual Bloom:* Bleichmar, *Visible Empire;* and O'Malley and Meyers, *Art of Natural History.*

10. It would be impractical (if not impossible) to list all the scholars who have worked on Edgeworth; those with a specific focus on her scientific interests include Ruth Watts, Nicole M. Wright, and Julia Douthwaite.
11. See Shteir, *Cultivating Women;* Fara, *Pandora's Breeches;* Kelley, *Clandestine Marriage;* and George, *Botany, Sexuality, and Women's Writing,* for earlier treatments of Wakefield and/or Jacson.
12. Aside from my own work, studies with an interest in the scientific work of Lennox and Haywood include Girten, "Unsexed Souls"; Italia, *Rise of Literary Journalism;* and Dorn, "Reading Women Reading History."
13. For a thoughtful study of the way that the garden served as a useful (if not unproblematic) site for educating children near the end of the eighteenth century, see Page and Smith, eds., *Women, Literature, and the Domesticated Landscape.*
14. See Saunders, *Picturing Plants;* and Daston and Galison, *Objectivity* (especially chap. 5), for more information on technical botanical illustration.
15. There are exceptions: Gibson's *Plant Contract* is a recent example of the move in contemporary scholarship to reexamine the critical significance of botanical artwork.
16. Current scholarship in the field of critical plant studies encompasses a range of topics from the investigation of plant intelligence to a political assessment of vegetal biopower, with significant recent work including Marder, *Plant-Thinking* and *Through Vegetal Being,* with Luce Irigaray; Nealon, *Plant Theory;* Kohn, *How Forests Think;* Mancuso, *Revolutionary Genius of Plants* and *Brilliant Green;* Gagliano, *Language of Plants;* and Hall, *Plants as Persons.*
17. Crucial work on this subject includes Batchelor and Kaplan, *Women and Material Culture;* Goggin and Tobin, *Women and Things* and *Women and the Material Culture;* and Pelling, *Duchess' Museum.*
18. The collection of essays credited with energizing Delany studies is Laird and Weisberg-Roberts, *Mrs. Delany and Her Circle.* See also Roberts, "Idleness Never Grew in My Soil"; Moore, "Queer Gardens" and *Sister Arts;* Tobin, *Duchess' Shells;* Peacock, *Paper Garden;* and Orr, *Mrs. Delany.*
19. We must thank Ann Shteir for bringing Moriarty to our attention, but she earns only a few paragraphs in *Cultivating Women, Cultivating Science.*
20. Letter to Thomas Cadell Jr. and William Davies, 1 August 1797, *Collected Letters of Charlotte Smith,* 283.
21. A theory first explicated in Shapin and Schaffer, *Leviathan and the Air-Pump,* the concept of scientific authority has since been complicated in a number of more recent studies, including Chico, *Experimental Imagination;* Coppola, *Theater of Experiment;* and Daston and Lunbeck, eds., *Histories of Scientific Observations.* Other interrogations of empiricism more broadly include Valenza, *Literature, Language;* Weiss-Smith, *Empiricist Devotions;* and Thompson, *Fictional Matter.*

22. See Linkin and Behrendt, eds., *Romanticism and Women Poets;* Behrendt, *British Women Poets;* Smith, *Romantic Women Writers;* Jacqueline M. Labbe, *Romantic Visualities* and *Charlotte Smith;* Mellor, *Romanticism and Gender;* Singer, *Romantic Vacancy;* and Sweet and Melnyk, *Felicia Hemans.*
23. For more information about the influence of geological discourses on poetry, see Goodman, "Conjectures on *Beachy Head*"; Bailes, *Questioning Science;* Heringman, *Romantic Rocks;* and Gidal, *Ossianic Unconformities.*
24. For a perceptive study of the shift in displays of scientific acumen between Smith and Hemans, see Bailes, "Linnaeus's Botanical Clocks."
25. Previous work that reads Smith's footnotes as supporting her own scientific aspirations includes Tayebi, "Undermining the Eighteenth-Century Pastoral"; Kelley, "Romantic Exemplarity"; Bailes, *Questioning Nature;* and Pascoe, "Female Botanists and the Poetry of Charlotte Smith."

1. Caterpillars in the Garden

1. Pearson, *Women's Reading in Britain,* 11; Fergus, "Women Readers," 156; Lawrence, *Women in England,* 165–75.
2. Pearson, 86.
3. Pope, "An Epistle to a Lady," 356, line 216.
4. Pearson, *Women's Reading in Britain,* 43.
5. Girten, "Unsexed Souls," 56.
6. Rousseau, "Science Books and Their Readers," 212.
7. Lawrence, *Women in England,* 179.
8. Adams, *Micrographia Illustrata,* iii.
9. Bailes, *Questioning Nature,* 8.
10. Gates and Shteir, *Natural Eloquence,* 4.
11. Haywood, *Female Spectator,* vol. 1, 19.
12. King, introduction to the *Female Spectator,* 1.
13. Spedding, "Measuring the Success," 198.
14. Wright and Newman, *Fair Philosopher,* 35.
15. Maurer, *Proposing Men;* and Hollis, "Eliza Haywood."
16. Haywood, *Female Spectator,* vol. 3, 104.
17. Smith, "Making as Knowing," 19.
18. For more information on the connections between domesticity and early science, especially with regards to women, see Hayden, ed., *New Science and Women's Literary Discourse.*
19. Haywood, *Female Spectator,* vol. 3, 81.
20. Haywood, 83.
21. Girten, "Unsexed Souls," 57–58.
22. Findlen, *Possessing Nature,* 258.
23. Haywood, *Female Spectator,* vol. 3, 90.
24. Haywood, 162–63.

25. Ross, *What Gardens Mean*, 129.
26. Haywood, *Female Spectator*, vol. 3, 163.
27. Haywood, 88.
28. Girten, "Unsexed Souls," 59.
29. See Rogers, *Feminism in Eighteenth-Century England;* Shevelow, *Women and Print Culture;* Spacks, *Selections from the* Female Spectator; Maurer, *Proposing Men;* and Miguel-Alfonso, "Social Conservatism, Aesthetic Education."
30. Prescott and Spencer, "Prattling, Tattling, and Knowing Everything," 47.
31. Bannet, "Haywood's Spectator," 98.
32. Girten, "Unsexed Souls," 61.
33. Haywood, *Female Spectator*, vol. 3, 85.
34. Cavendish, *Blazing World and Other Writings*, 141–43.
35. Haywood, *Female Spectator*, vol. 3, 161.
36. Girten, "Unsexed Souls," 60–61.
37. Haywood, *Female Spectator*, vol. 3, 84.
38. Haywood, 162.
39. Haywood, 165.
40. "Insects" was a capacious term in the eighteenth century that also encompassed a number of mollusks, cnidaria, and other invertebrates. Hence, while snails are not arthropods in modern classification systems (and would therefore not be considered insects today), Haywood is using contemporary terminology accurately.
41. Haywood, *Female Spectator*, vol. 3, 162.
42. Ogilvie, *Science of Describing*, 212–14.
43. Haywood, *Female Spectator*, vol. 3, 162.
44. Haywood, 162.
45. Haywood, 164.
46. Haywood, 164.
47. *Snail*, n.p.
48. Haywood, *Female Spectator*, vol. 3, 165.
49. Haywood, 164. This is especially interesting given that most snails are hermaphrodites.
50. Snook, "Beautifying Part of Physic," 16.
51. Haywood, *Female Spectator*, vol. 3, 164.
52. Baster, "Dissertation on the Worms," 279.
53. Bonnet, "Some New Observations upon Insects," 460.
54. Haywood, *Female Spectator*, vol. 3, 162.
55. Terrall, "Following Insects Around," 574.
56. Terrall, 574.
57. In this respect, I disagree with Gordon's argument that Arabella's good qualities exist not because of, but in spite of, her quixotic reading habit. See *Practice of Quixotism*.

58. The appropriate exercise of wit was also a popular subject with theatre critics, especially John Dennis, Jeremy Collier, and Oliver Goldsmith: Langbauer, *Women and Romance,* 69.
59. Lennox, *Female Quixote,* 266.
60. Lennox, 328–29.
61. Palo, "Good Effects of a Whimsical Study," 209.
62. See Raven, Small, and Tadmor, eds., *Practice and Representation of Reading;* Vickery, *Gentleman's Daughter;* and Pearson, *Women's Reading in Britain.*
63. Lennox, *Female Quixote,* 328.
64. Gordon, "Space of Romance," 510.
65. Lennox, *Female Quixote,* 373, 266.
66. Lennox, 394n.
67. Mack, "Quixotic Ethnography," 207.
68. Mack, 204–6.
69. Mack, 197.
70. Lennox, *Female Quixote,* 35.
71. Palo, "Whimsical Study," 220.
72. Vickery, *Gentleman's Daughter,* 258; Pearson, *Women's Reading in Britain,* 52.
73. December 1760 was skipped due to some sort of printing error, according to Norbert Schürer. See introduction to Lennox's *Sophia,* 32.
74. Carlile, *Charlotte Lennox,* 173.
75. Ballaster et al., *Women's Worlds,* 44. By comparison, as a separate novel *Sophia* cost more than five shillings and scientific textbooks could cost around four to ten shillings at the lowest price point. Schürer, introduction to Lennox's *Sophia,* 48; Rousseau, "Science Books and Their Readers," 233.
76. Carlile, *Charlotte Lennox,* 171.
77. For an insightful reading of the Trifler that emphasizes how significant her role was in the periodical, see Carlile, "Bear the Steady Ray of Reason."
78. Bataille, *Writing Life of Hugh Kelly,* 5.
79. For more information about how involved Strahan, Johnson, and Richardson were in promoting Lennox's work, see Schürer, ed., *Correspondence and Miscellaneous Documents.*
80. Carlile, *Charlotte Lennox,* 200–202.
81. Lennox, *Lady's Museum,* 4.
82. Lorenzo-Modia, "Education for Women in the Eighteenth-Century Periodical," 204–5. This is actually the first recorded English translation of the treatise, printed only two years after it was published in French.
83. Jouslin, "Conservative or Reformer?," 71.
84. Jouslin, 49.
85. Italia, *Rise of Literary Journalism,* 3.
86. Lennox, *Lady's Museum,* 261–68.
87. Lennox, 130.

88. Jouslin, "Conservative or Reformer?," 66.
89. Shevelow, "Preview," 182.
90. Dorn, "Reading Women Reading History," 20.
91. Lennox, *Lady's Museum*, 629.
92. For more on the radical intellectual potential of the history essays in the *Lady's Museum*, see my article "Constructing Women's History."
93. Lennox, *Lady's Museum*, 129.
94. Lennox, 11.
95. Lennox, 297.
96. Lennox, 130.
97. Lennox, 134.
98. Lennox, 136.
99. Lennox, 136.
100. Murray, *British Garden*, vi.
101. Shteir, "Botanical Dialogues," 311.
102. Lennox, *Lady's Museum*, 467, 633.
103. Lennox, 235, 237, 130.
104. Lennox, 229–40.
105. Lennox, 467–68.
106. Lennox, 469.
107. A ghastly exhibition of entomological parasitism that disturbed Charles Darwin's own belief in natural theology. See Mahood, *Poet as Botanist*, 79. Lennox, however, seems unfazed.
108. Lennox, *Lady's Museum*, 235. While zoology would be considered unsuitable for women to study by the early years of the nineteenth century, at the time Lennox was publishing her periodical it was acceptable within vague limits of propriety.
109. Lennox, 136.
110. Lennox, 137.
111. Lennox, 139.
112. Lennox, 137–38.
113. Lennox, 136–37.
114. Lennox, 133.
115. Carlile, *Charlotte Lennox*, 176.
116. Lennox, *Lady's Museum*, 862.
117. Lennox, 858.
118. Lennox, 857.
119. Spacks, *Selections from the* Female Spectator, xii. An epithet that is also likely untrue.
120. Haywood, *Female Spectator*, vol. 3, 175.

2. From Native Blooms to Monster Plants

1. Smith, *Lachesis Lapponica*, 188.
2. Bewell, "On the Banks," 177.
3. Kelley, *Clandestine Marriage*, 93.
4. La Mettrie, *Machine Man*, 79. The use of plant descriptors to convey pornographic imagery was hardly uncommon; the mimosa in particular was utilized in pornographic texts to refer to the penis, and fictional plants like the "frutex vulvaria" to refer to female genitalia. See, for example, John Cleland's *Memoirs of a Woman of Pleasure* (1748), and the pornographic treatise *The Natural History of the Frutex Vulvaria* by the pseudonymous Philogynes Clitorides (1732).
5. La Mettrie, *Machine Man*, 80.
6. Schiebinger, "Private Life of Plants," 136.
7. Koerner, *Linnaeus*, 15.
8. George, *Botany, Sexuality, and Women's Writing*, 25.
9. Lynch, "Young Ladies Are Delicate Plants," 699.
10. Lynch, 699.
11. Quoted in Wollstonecraft, *Vindication*, 123.
12. Quoted in Wollstonecraft, 123.
13. Wollstonecraft, 98.
14. Bewell, "Jacobin Plants," 138.
15. George, *Botany, Sexuality, and Women's Writing*, 16.
16. For more information on eighteenth-century flower arrangements, see Mary Rose Blacker's *Flora Domestica*, especially chap. 3, for its relevance to the time when Jacson and Wakefield were writing.
17. Shteir, *Cultivating Women*, 3–4.
18. Shteir, 4.
19. Kelley, *Clandestine Marriage*, 103.
20. Shteir, "Botanical Dialogues," 311.
21. Myers, "Impeccable Governesses," 34.
22. Teute, "Loves of the Plants," 332.
23. Polwhele, *Unsex'd Females*, 10.
24. Oddly enough, Polwhele approved of *The Botanic Garden* for its "glowing" and "sublime" poetic merits, arguing in a footnote to *The Unsex'd Females* that "science may sometimes be brought forward, not unhappily, under the conduct of imagination." Polwhele's concern with the text, then, is rooted in his disapproval of those women who might follow the radical (botanical) politics of women like Wollstonecraft, Mary Robinson, and Mary Hays. See Polwhele, *Unsex'd Females*, 4–5.
25. Jacson, *Botanical Dialogues*, 65, 124, 140.
26. See 179n6 and 180n2 in Knowles and Horricks, eds., *Charlotte Smith*.

27. For more on botany and empire, see chap. 4.
28. Schiebinger, *Plants and Empire*, 11.
29. See Bewell, "Romanticism and Colonial Natural History," 16.
30. Bewell, "Erasmus Darwin's Cosmopolitan Nature," 32.
31. Darwin, *Loves of the Plants*, canto 1, line 65. Curcuma is only part of the plant's name—Darwin does not get more specific. In all likelihood, he means Curcuma longa (the source of turmeric).
32. For example, it can be found as a treatment for irregular bleeding in the pseudonymously authored *Every Lady Her Own Physician*, 28.
33. See Brookes, *New and Accurate System*, 21.
34. Darwin, *Loves of the Plants*, canto 1, lines 45, 119, 204.
35. Darwin, canto 1, lines 264 and 282.
36. Acclimatization is actually a very complex process that could entail any number of tactics and approaches. It was pursued both professionally at nurseries and botanical gardens and at personal homes of those with enough resources. For more information about eighteenth-century acclimatization gardens, see Harris, *Planting Paradise*, primarily chap. 6.
37. Bewell, "Erasmus Darwin's Cosmopolitan Nature," 22.
38. Bewell, 27.
39. For an extended study of monstrous female plant life in *The Loves of the Plants* (which I do not discuss in this chapter), see Heckendorn Cook, "'Perfect' Flowers."
40. "Plantago rosea" is not an accepted species designation; Darwin is likely referring to *Plantago major*, also colloquially known as the rose plantain for its blossom-like shape.
41. Darwin, *Loves of the Plants*, canto 1, lines 77–84. While "Desdemona" is an accepted synonym for a plant, that plant (genus *Basistemon*) was not discovered until the mid-nineteenth century in South America. Darwin is thus imagining not a plant woman in this scenario, but a human woman-turned-plant.
42. Lennox, *Shakespear Illustrated*, vol. 1, 130.
43. Lee, *Introduction to Botany*, 231.
44. Wakefield, *Introduction to Botany*, letter 4, 19–20.
45. A fascinating and extensive series of observations about the poisonous Upas Tree (*Antiaris toxicaria*) can be found in canto 3, to name only the most famous example.
46. Shirk, "Contributions to Botany," 296.
47. Wakefield, preface to *Introduction to Botany*, n.p.
48. In this respect, I disagree with Shteir and Marina Benjamin, who propose that Wakefield's use of religious themes was little more than perfunctory, or a ploy to avoid censorship. As evidenced in the works of Lennox and Delany, women often evoked religion in the context of natural history and botany but rarely to a significant extent. See *Cultivating Women*, 85, 39.

49. Wakefield, *Introduction to Botany,* letter 25, 153.
50. Foucault, *Order of Things,* 135.
51. Wakefield, preface to *Introduction to Botany,* n.p.
52. Wakefield, n.p.
53. Wakefield, n.p.
54. Wakefield, *Introduction to Botany,* letter 1, 2.
55. Wakefield, letter 2, 8.
56. Wakefield, letter 10, 58.
57. Wakefield, letter 3, 10.
58. Wakefield, letter 2, 7.
59. Wakefield, letter 3, 10.
60. Wakefield, preface to *Introduction to Botany,* n.p.
61. Shirk, "Contributions to Botany," 300, 302.
62. Fara, *Pandora's Breeches,* 203.
63. Wakefield, *Introduction to Botany,* letter 5, 23.
64. Wakefield, letter 23, 136.
65. Wakefield, letter 4, 18.
66. Wakefield, letter 8, 43.
67. Wakefield, letter 1, 2.
68. Wakefield, letter 15, 88.
69. Wakefield, letter 9, 47.
70. Felicia has in fact taken up botany expressly for the purpose of combatting depression; she writes, "Even the approach of spring . . . affords me but little pleasure; my kind mother, ever attentive to my happiness, concurs with my governess in checking this depression of spirits, and insists upon my having recourse to some interesting employment that shall amuse me . . . my fondness for flowers has induced my mother to propose Botany." Wakefield, letter 1, 2.
71. Public pleasure gardens and town gardens also become increasingly popular in the eighteenth century, providing women access to garden spaces (not unproblematically) even if they didn't have the ability to cultivate their own gardens. See Bending, *Green Retreats,* 3.
72. Shteir, "Botanical Dialogues," 308.
73. Jacson, *Botanical Dialogues,* advertisement, n.p.
74. Jacson, *Botanical Dialogues,* 2.
75. Jacson, 4.
76. Jacson, vii.
77. Jacson, 5.
78. For one example, Hortensia finds this to be the only flaw in William Curtis's otherwise excellent *Botanical Magazine,* which was popular, affordable at one shilling per issue, and gorgeously illustrated. See Jacson, 65.
79. Jacson, 65–66.
80. Jacson, 71.

81. Jacson, 71.
82. Jacson, 274.
83. Uglow, *Lunar Men*, 380–81.
84. Jacson, *Botanical Dialogues*, 3–4.
85. Jacson, 66.
86. Jacson, 35.
87. Jacson, 238.
88. Jacson, 239.
89. Jacson, 55.
90. See Mathias, *Pursuits of Literature*.
91. Jacson, *Botanical Dialogues*, 2.
92. Jacson, 13, 21, 24.
93. Rousseau, *Elements of Botany*, xi–xii.
94. Jacson, *Botanical Dialogues*, 229.
95. Jacson, 213.
96. Jacson, 33.
97. Murray, *British Garden*, 284.
98. Laird, *Natural History of English Gardening*, 101.
99. Lynch, "Young Ladies Are Delicate Plants," 697.
100. Pope, "Epistle to a Lady," 351.
101. Swift, "Lady's Dressing Room," 8.
102. Lynch, "Young Ladies Are Delicate Plants," 704.
103. Miller, *Vegetative Soul*, 10.
104. Wakefield, *Introduction to Botany*, letter 2, 4–5.
105. Delaporte, *Nature's Second Kingdom*, 2.
106. Wakefield, *Introduction to Botany*, letter 2, 5.
107. Barbauld, "To a Lady," lines 10–15.
108. Wakefield, *Introduction to Botany*, letter 2, 9.
109. Wakefield, letter 2, 9.
110. Wakefield, letter 6, 32.
111. Wakefield, letter 7, 32.
112. Jacson, *Botanical Dialogues*, 329–30.
113. Lee, *Introduction to Botany*, 231.
114. Miller, "Lives of the Monster Plants," 462.
115. See Jacson, *Botanical Dialogues*, 259, for an example.
116. Jacson, 278.
117. For more information on the different axes which separated humans and plants in eighteenth-century thought, see Delaporte, *Nature's Second Kingdom*.
118. Cohen, "Monster Culture," 6.
119. Keetley, "Introduction," 5.
120. Jacson, *Botanical Dialogues*, 43.
121. We might think, for one, of Frances Burney's *Cecilia* and the character Bel-

field, whose indolence and wastrel ways impoverish his previously well-off parents.

122. La Mettrie, *Machine Man,* 84. "Without feeling it"—a curious sentiment to appear in an essay that also imagines plant vaginas.
123. Delaporte, *Nature's Second Kingdom,* 151.
124. Houle, "Animal, Vegetable, Mineral," 100–103. For more information on plant senses and intelligence, see Mancuso and Viola, *Brilliant Green.*
125. Darwin, *Phytologia,* 132–33.
126. Darwin, 135.
127. Marder, *Plant-Thinking,* 12.
128. Miller, "Lives of Monster Plants," 462.
129. Keetley, "Introduction," 9.
130. Nealon, *Plant Theory,* 11.

3. Pedagogies and Possibilities

1. This exchange may be inspired by the relationship between Richard Lovell Edgeworth and Thomas Day. See Butler, *Maria Edgeworth,* 149; and Ó Gallchoir, "Gender, Nation, and Revolution," 204.
2. Edgeworth, *Letters for Literary Ladies,* 1.
3. Edgeworth, 20.
4. Edgeworth, 20.
5. Narain, "Prescription of Letters," 268–69.
6. Edgeworth, "Note on the Text," in *Letters for Literary Ladies,* xxvii.
7. Belanger, "Publishers and Writers," 5–7; Shteir, *Cultivating Women,* 3–4.
8. Armstrong, "Rise of the Domestic Woman," 100.
9. Edgeworth, *Practical Education,* 173.
10. Edgeworth, 311.
11. Chapone, *Letter to a New-Married Lady,* 16–17.
12. Edgeworth, *Practical Education,* 311–12.
13. Phillips, *Scientific Lady,* 100.
14. Stafford, *English Feminists,* 163.
15. Narain, "Prescription of Letters," 283.
16. Edgeworth, *Letters for Literary Ladies,* 20.
17. Edgeworth, 40. In this, Julia's letter is echoing the first section of *Letters,* in which the misogynist "Gentleman" of the first letter cites Molière's *Les Femmes Savantes* as a reason to avoid educating daughters in the sciences (see Edgeworth, 11).
18. Edgeworth, 37.
19. Edgeworth, 40.
20. Edgeworth, 56.
21. Edgeworth, 48.
22. Quoted in Stafford, *English Feminists,* 10.

23. Keen, *Crisis of Literature in the 1790s,* 194.
24. Edgeworth, *Letters for Literary Ladies,* 20.
25. Edgeworth, 29.
26. Butler, *Maria Edgeworth;* Gonda, *Reading Daughter's Fictions,* 273.
27. Butler, 31–33.
28. For more on the Lunar Society, see Uglow, *Lunar Men;* with respect to Maria Edgeworth's work in particular, see Chandler, "Edgeworth and the Lunar Enlightenment."
29. Watts, "Whose Knowledge?," 290.
30. Edgeworth, *Practical Education,* 278–82; Edgeworth, *Letters for Literary Ladies,* 21.
31. Edgeworth, *Letters for Literary Ladies,* 21.
32. Edgeworth, *Practical Education,* 196; Edgeworth, "Good French Governess," 300.
33. Edgeworth, "Good French Governess," 301.
34. Edgeworth "Forester," 182.
35. Edgeworth, 241.
36. Edgeworth, *Practical Education,* 312.
37. Edgeworth, "Angelina," 278–79.
38. Edgeworth, "Good French Governess," 293; Edgeworth, *Belinda,* 235.
39. Quoted in Butler, *Maria Edgeworth,* 83.
40. Edgeworth, *Letters for Literary Ladies,* 21.
41. Weiss, "Extraordinary Ordinary Belinda," 444.
42. See Narain, "Not the Angel"; Toal, "Control Experiment"; Watts, "Whose Knowledge?"; and Douthwaite, "Experimental Child-Rearing."
43. Manly, introduction to Edgeworth's *Practical Education,* xi.
44. See Douthwaite, "Experimental Childrearing"; and Coppola, *Theater of Experiment.*
45. Wright, "Opening the Phosphoric 'Envelope,'" 511; Edgeworth, *Belinda,* 222.
46. Wright, 511.
47. See Scantlebury and Murphy, "Maria Edgeworth."
48. Holmes, *Age of Wonder,* 264.
49. Edgeworth, *Practical Education,* 195.
50. Reeve, *Progress of Romance,* 111.
51. Edgeworth, *Practical Education,* 22, 27.
52. Edgeworth, "Good French Governess," 292.
53. Edgeworth, 293.
54. Edgeworth, *Practical Education,* 27.
55. Turner, *Scientific Instruments,* 95–96.
56. Edgeworth, *Letters for Literary Ladies,* 20. This is written in the voice of the defensive father, whose advocacy for female education relies in part upon its domestic utility.

57. Kirkpatrick, "Limits of Liberal Feminism," 74, 77.
58. Douthwaite, *Wild Girl,* 184.
59. Edgeworth, *Practical Education,* 5.
60. Edgeworth, *Belinda,* 215.
61. Edgeworth, 215.
62. Edgeworth, 216.
63. Edgeworth, 216.
64. Quoted in Butler, *Maria Edgeworth,* 212–13.
65. Edgeworth, *Belinda,* 99.
66. Wright, "Opening the Phosphoric 'Envelope,'" 533.
67. Edgeworth, *Belinda,* 99.
68. See, for example, "Part of a Letter from Abbe Nollet, of the Royal Academy of Sciences at Paris, and F. R. S. to Martin Folkes Esq . . . concerning Electricity."
69. Kirkpatrick, *Belinda,* note to page 99.
70. Nollet, *Leçons,* 415.
71. Nollet, *Leçons,* 420.
72. Nollet, *Leçons,* 421–22.
73. Letter to Richard Lovell Edgeworth, 24 November 1815, in *Maria Edgeworth's Letters from Ireland,* 187.
74. Edgeworth, *Belinda,* 167.
75. MacFayden, "Lady Delacour's Library."
76. For more on the cultural associations with goldfish in the eighteenth century, see Kaul, "Why Selima Drowns," 227.
77. Edgeworth, *Belinda,* 172.
78. Edgeworth, 162.
79. Edgeworth, 120.
80. Cheater, "Collectors of Nature's Curiosities," 170; Jackson, *Sarah Stone,* 37.
81. Cheater, 170–71; Kim Sloan, *Noble Art,* 228.
82. See *Companion to the Museum,* 1–2, 25–27.
83. Quoted in Cheater, "Collectors of Nature's Curiosities," 171.
84. Jackson, *Sarah Stone,* 42, 44, 48.
85. Cheater, "Collectors of Nature's Curiosities," 173–74; Jackson, *Sarah Stone,* 50–52.
86. Letter to Lucy Edgeworth, 18 March 1815, in *Letters from Ireland,* 192.
87. Jackson, *Sarah Stone,* 52.
88. Cheater, "Collectors of Nature's Curiosities," 179n23; Sloan, *Noble Art,* 228.
89. Cheater, 173.
90. Kim Sloan, *Noble Art,* 226; Jackson, *Sarah Stone,* 48.
91. Edgeworth, *Belinda,* 162.
92. For a full account of Bewick's career, see Uglow, *Nature's Engraver.*
93. Uglow, 43.
94. Beilby, *History of British Birds,* 166.

95. Letter to Sophy Ruxton, 26 February 1805, in *Letters from Ireland*, 91.
96. Edgeworth, *Belinda*, 235.
97. "Lemna," Plants of the World Online, accessed May 18, 2021, http://powo.science.kew.org/taxon/urn:lsid:ipni.org:names:23865-1.
98. Edgeworth, *Belinda*, 285. The quotation marks in Belinda's last comment are reproduced exactly from the text, but I am unable to find a source for the quotation.
99. Hilton, *Shaping of the Nation's Young*, 126n50. However, their authorship was apparently not a secret among family circles; see MacDonald, ed., *Education of the Heart*, 52.
100. See MacDonald, 49–51, 114, 119, 124–25, for only a few examples; see also Hilton, *Shaping of the Nation's Young*, 126.
101. Letter to Sophy Ruxton, 22 January 1794; and letter to Letty Ruxton, [n.d.] 1794, in *Letters from Ireland*, 37, 45.
102. See, for example, letter to Aunt Ruxton, September 1807, in *Letters from Ireland*, 110; and MacDonald, ed., *Education of the Heart*, 60–62, 70.
103. MacDonald, 72.
104. B. D. Jackson, "Edgeworth, Michael Pakenham (1812–1881), Botanist and East India Company Servant," Oxford Dictionary of National Biography, September 23, 2004, https://www.oxforddnb.com. Pakenham also sent a number of seeds from India to Maria in Ireland, some of which she forwarded on to Lazarus. MacDonald, ed., *Education of the Heart*, 285.
105. Edgeworth, *Practical Education*, 236.
106. Edgeworth, 239.
107. Edgeworth, 251.
108. Edgeworth, 278; Edgeworth, *Belinda*, 222.
109. See, for example, Tobias Smollett, *The History and Adventures of an Atom* (1769); Anonymous, *The Adventures of an Ostrich Feather of Quality* (1812); and Eliza Haywood, *The Anti-Pamela; or Feign'd Innocence Detected* (1741).
110. While not all critics agree that Delacour has been redeemed by the conclusion of the novel, I find the evidence in favor of her rehabilitation most compelling.
111. King, *Bloom*, 59.
112. MacFayden, "Lady Delacour's Library," 426.
113. Edgeworth, *Belinda*, 62.
114. Edgeworth, 63.
115. Edgeworth, *Practical Education*, 311–12.
116. "Aloe vera," Plants of the World Online, accessed May 21, 2021, http://powo.science.kew.org/taxon/urn:lsid:ipni.org:names:530017-1.
117. Abercrombie, *Every Man His Own Gardener*, 740–41.
118. Miller, "Aloe," in *Gardeners Dictionary*, n.p.
119. Miller, n.p.
120. Miller, n.p.

121. Edgeworth, *Belinda*, 221.
122. Edgeworth, 222.
123. Blair, *Pharmaco-Botanalogia*, 58.
124. Blair, 58.
125. Edgeworth, *Belinda*, 63.
126. Edgeworth, 69.
127. Edgeworth, 55–58.
128. See, for example, Kowaleski-Wallace, *Their Fathers' Daughters;* Klein, "Bosom Friends"; Montwieler, "Reading Disease"; and Greenfield, "Abroad and at Home."
129. Edgeworth, *Belinda*, 292.
130. Edgeworth, 290.
131. Edgeworth, 292.
132. "Prunus cerasus," Plants of the World Online, May 21, 2021, http://powo.science.kew.org/taxon/urn:lsid:ipni.org:names:729574-1.
133. Edgeworth, "Forester," 205.
134. Edgeworth, *Belinda*, 308.
135. Edgeworth, 287. I am classifying the "Provence rose" as a variety of *Rosa centifolia* on the basis of several eighteenth-century gardening dictionaries, including Abercrombie's *Gardener's Pocket Dictionary*, as well as contemporary gardening resources.
136. Miller, "Rosa," in *Gardener's Dictionary*, n.p.
137. See Linnaeus, *Species Plantarum*, 175, 492; and Ray, *Historia Plantarum*, 32.
138. Cullen, *Treatise of the Materia Medica*, 212; Aikin, *Manual of Materia Medica*, 144.
139. Edgeworth, *Belinda*, 270–71.
140. Edgeworth, 50.

4. Sketching Vegetality

1. See Shapin and Schaffer, *Leviathan and the Air-Pump*. Their account of eighteenth-century empiricism has since been further explored and complicated by many scholars, including Weiss-Smith, Valenza, and Chico.
2. Bending, *Green Retreats*, 1.
3. Cunningham, "Culture of Gardens," 50.
4. Blunt and Raphael, *Illustrated Herbal*, 165.
5. Daston and Galison, *Objectivity*, 86.
6. Blackwell, *Curious Herbal*, plate 45.
7. Moriarty, *Viridarium*, plate 44.
8. Tobin, "Imperial Designs," 268.
9. Scott, *Millenium Hall*, 118.
10. Brown, *Sisters of St. Gothard*, 161–62.
11. Smith, *Marchmont*, vol. 3, 174.

12. Chapman, *Eighteenth-Century Women Artists*, 11, 105. Along with Angelica Kauffman, she was one of the only female members of the Royal Academy for a number of years.
13. Chapman, 105.
14. Adams, *Cyclopædia of Female Biography*, 471.
15. King, *Botanical Illustration*, 10; Blunt and Stearn, *Art of Botanical Illustration*, 153.
16. Cheston, "Elizabeth Blackwell," 25.
17. For example, we know that Sophie Cuvier sketched animal portraits for her father Georges, just as Joseph Dalton Hooker's daughter Harriet painted plants for her father's journal. See Daston and Galison, *Objectivity*, 89.
18. Kramer, *Women of Flowers*, 35.
19. "Address," *New Treatise on Flower Painting*, n.p.
20. Davis, *Women on the Margins*, 143.
21. Blackwell, *Curious Herbal*, descriptions of plates 387 and 388.
22. Tobin, "Imperial Designs," 266.
23. Nyberg, "Linnaeus's Apostles," 88.
24. Madge, "Elizabeth Blackwell," n.p.
25. Blackwell, *Curious Herbal*, dedication to vol. 2, written in 1737.
26. Blackwell, preface to *Curious Herbal*, n.p.
27. Blackwell, preface to *Curious Herbal*, n.p.
28. "Register of Books."
29. Clayton, *English Female Artists*, 98.
30. Pulteney, *Progress of Botany*, 251.
31. Madge, "Elizabeth Blackwell," n.p.
32. Blunt and Raphael, *Illustrated Herbal*, 176.
33. *Plant*, 64.
34. Clayton, *English Female Artists*, 95.
35. Along the way, he wrote an agricultural treatise on how to drain wet or clay-dominant soil, which apparently attracted the attention of the Swedish Ambassador. See Blackwell, *New Method*.
36. *Genuine Copy*, 16.
37. Koerner, *Linnaeus*, 34.
38. Koerner, 39.
39. Koerner, 47.
40. Blackwell, *Curious Herbal*, prefatory material, n.p.
41. Blunt and Stearn, *Art of Botanical Illustration*, 153.
42. Leong, *Recipes and Everyday Knowledge*, 7.
43. Leong, 56.
44. Blackwell, *Curious Herbal*, description of plate 374.
45. Blackwell, description of plate 338.
46. Blackwell, descriptions of plates 341–44.
47. Shaw and Nodder, *Naturalist's Miscellany*, vol. 3, plate 82.

48. Kramer, *Women of Flowers*, 146.
49. Moriarty, dedication to *Fifty Plates*. This second, retitled edition was consulted for the production of this study.
50. Moriarty, *Brighton in an Uproar*, 136.
51. Moriarty, 136.
52. Moriarty, preface to *Fifty Plates*, v–vi.
53. Moriarty, vi–vii.
54. Moriarty, vii.
55. Britten, "Mrs. Moriarty's 'Viridarium,'" 53.
56. Daston and Galison, *Objectivity*, 99.
57. Moriarty, *Fifty Plates*, description of plate 3.
58. Moriarty, *Fifty Plates*, description of plate 4.
59. Bailes, *Questioning Nature*, 94.
60. See Page and Smith, eds., *Women, Literature, and the Domesticated Landscape*.
61. Ross, *What Gardens Mean*, note to image 3.
62. Ross, 80.
63. Page and Smith, *Domesticated Landscape*, 1.
64. Alaimo, *Undomesticated Ground*, 20.
65. Bending, *Green Retreats*, 5.
66. Bending, 5.
67. Lynch, "Young Ladies Are Delicate Plants," 692.
68. Lynch, 693.
69. See, for example, work by Stacy Alaimo, Vicki Kirby, Catriona Mortimer-Sandilands, Val Plumwood, Donna Haraway, Karen Barad, and Lara Farina. I have also found the work of Jeffrey Jerome Cohen, Lowell Duckert, Tobias Menely, Vin Nardizzi, and Eric Gidal to be valuable in theorizing "nature."
70. Bewell, *Natures in Translation*, 9–21.

5. Collecting and Creating

1. Delany, *Autobiography and Correspondence*, series 2, vol. 1, 469.
2. The most recent sustained studies can be found in the excellent collection *Mrs. Delany and Her Circle*, edited by Laird and Weisberg-Roberts. Recent biographies include Peacock's *Paper Garden* and Clarissa Campbell Orr's *Mrs. Delany, a Life*.
3. See, for example, Pyle, "Artist or Artistic?," which ultimately ends on an indecisive middle ground.
4. Fennetaux, "Female Crafts," 100.
5. Tobin, *Duchess's Shells*, 99.
6. Schiebinger, *Nature's Body*, 36.
7. Moore, "Queer Gardens," 57–58.

8. Haut, "Reading Flora," 245–46.
9. Smith, "Making as Knowing," 19.
10. Gere and Vaizey, *Great Women Collectors*, 87.
11. Tobin, *Duchess's Shells*, 40.
12. Fara, *Sex, Botany, and Empire*, 12.
13. Delany, *Autobiography and Correspondence*, 2.1:384.
14. Delany, 2.1:384.
15. Laird and Weisberg-Roberts, *Mrs. Delany and Her Circle*, 1.
16. Delany, *Autobiography and Correspondence*, 1.3:544.
17. Pelling, "Collecting the World," 1.
18. With respect to cloth production, see Weisberg-Roberts, "Between Trade and Science."
19. Delany, *Autobiography and Correspondence*, 2.1:18–19.
20. This, in part, formed the basis of her persistent disapproval of Rousseau despite the friendly relationships between other members of her family and the French philosophe. A series of letters in 1767 cautions her niece against closer acquaintance with her then neighbor: Delany, 2.1:80.
21. Delany, 1.1:484.
22. Vickery, "Theory and Practice," 107.
23. See, for example, Delany, *Autobiography and Correspondence*, 2.1:281.
24. Delany, 1.3:39. The reference to "Irish stich" would be a compliment to the regularity of her mosaic.
25. Delany, 2.1:281.
26. Moore, "Queer Gardens," 61.
27. Delany, *Autobiography and Correspondence*, 1.3:544.
28. Moore, "Queer Gardens," 61.
29. Delany, *Autobiography and Correspondence*, 2.2:345.
30. Delany, 2.2:345.
31. Delany, 2.2:563.
32. Delany, 2.2:442.
33. Delany, 2.2:433.
34. Delany, 2.2:39–40.
35. Delany, 2.1:277.
36. Delany, 2.1:405.
37. Delany, 2.1:410.
38. Delany, 2.1:243–44.
39. Delany, 2.1:374.
40. Delany, 2.2:47.
41. Delany, 2.2:47.
42. Delany, 2.2:47.
43. Delany, 2.2:410.
44. Delany, 2.1:238.
45. Zytaruk, "Mary Delany," 141.

46. *Catalogue of the Portland Museum,* iii. See also Sloboda, "Displaying Materials."
47. *Catalogue of the Portland Museum,* iii.
48. *Catalogue of the Portland Museum,* iii.
49. Delany, *Autobiography and Correspondence,* 2.1:274, 2.1:170.
50. Delany, 2.1:337.
51. Delany, 2.2:366.
52. Ford, "Progress in Plants," 220.
53. Ford, 207, 211.
54. Delany, *Autobiography and Correspondence,* 2.1:479.
55. Delany, 2.2:365.
56. Delany, 2.1:243.
57. Delany, 2.1:243.
58. Delany, 2.1:512.
59. Delany, 2.1:512.
60. Delany, 2.2:442.
61. Fara, *Sex, Botany, and Empire,* 32.
62. Koerner, "Carl Linnaeus," 148.
63. Edmonson, "Novelty in Nomenclature," 190.
64. Moore, "Queer Gardens," 61.
65. Darwin, *Loves of Plants,* lines 155–58.
66. Darwin, lines 163–64.
67. For more on the biological possibilities of plant sentience, see Mancuso and Viola, *Brilliant Green;* Baluška, Mancuso, and Volkmann, eds., *Communication in Plants;* multiple studies by Braam; and Hall, *Plants as Persons.* For a collection of essays that incorporates biological perspectives into cultural ones, see Heyd, ed., *Recognizing the Autonomy of Nature.* Mabey's *Cabaret of Plants* is an excellent cultural overview.
68. Delany, *Autobiography and Correspondence,* 2.2:223.
69. Delany, 2.2:223.
70. Delany, 2.2:223.
71. Wirt, *Flora's Dictionary,* 54.
72. Saunders, *Picturing Plants,* 96.
73. Saunders, 12.
74. Ford, "Progress in Plants," 220.
75. Delany, *Autobiography and Correspondence,* 2.2:421.
76. Delany, 2.2:326.
77. Delany, 2.2:424.
78. Delany, 2.1:469.
79. Delany, 2.2:215.
80. Edmonson, "Novelty in Nomenclature," 188.
81. Hayden, *Mrs. Delany,* 15.
82. Sloan, *Noble Art,* 46.

83. Delany, *Autobiography and Correspondence*, 2.1:173.
84. Quoted in Laird and Weisberg-Roberts, eds., *Mrs. Delany and Her Circle*, 10.
85. Delany, *Autobiography and Correspondence*, 2.3:95.
86. Reeder, for example, specifies that, in some cases, the sexual parts of the plant were simply too small to represent with paper, although the significance of this to a critical assessment of their value seems limited. See Reeder, "Paper Mosaick," 228.
87. Delany, *Autobiography and Correspondence*, 2.2:249 and 372.
88. Delany, 2.2:316.
89. Delany, 2.2:416.
90. Blunt and Stearn, *Art of Botanical Illustration*, 148.
91. Gosse, "Of Paper Flowers," 508.
92. Reeder, "Paper Mosaick," 231.
93. Lisa Moore raises the possibility of the German connection because the Duchess of Portland owned some of Dietzsch's work. See Moore, *Sister Arts*, 42.
94. To be more specific, 151 of the 985 extant mosaics are catalogued by the British Museum as containing plant matter.
95. Kippis, "Delany, Mary," 91.
96. Gowrley, "Craft(ing) Narratives," 80.
97. Delany, *Autobiography and Correspondence*, 2.2:487.
98. Delany, 2.2:487–89. The uncertainty about the date is Llanover's.
99. Delany, 2.2:481.
100. Delany, 2.2:48.
101. Delany, 2.2:48.
102. Delany, 2.2:443.
103. Linnaean Collections, accessed May 20, 2021, https://www.linnean.org/research-collections/linnaean-collections.
104. Delany, *Autobiography and Correspondence*, 2.2:443.
105. Delany, 2.2:443.
106. Delany, 2.3:95.
107. Delany, 2.3:98.
108. George, *Botany, Sexuality, and Women's Writing*, 32.
109. Delany, *Autobiography and Correspondence*, 2.3:98.
110. Delany, 2.1:173.
111. Delany, 2.3:4.

Coda

1. Tayebi, "Undermining the Eighteenth-Century Pastoral," 133.
2. Kelley, "Romantic Exemplarity," 224.
3. Kranz, "Language of Flowers," 195.

4. Kelley, "Romantic Exemplarity," 231. See also Bailes, "Linnaeus's Botanical Clocks," 238.
5. Marder, *Plant-Thinking*, 5.
6. Marder, 5.
7. Many have noted that the popularity of the Linnaean system endured in England so long in part because his works were among the few translated in English and in part because of the widespread English hostility toward French intellectual work.
8. Bailes, "Linnaeus's Botanical Clocks," 244.
9. Smith, "Flora," lines 14 and 226. Hereafter cited parenthetically by line numbers.
10. Tayebi, "Undermining the Eighteenth-Century Pastoral," 133.
11. Pascoe, "Female Botanists," 195.
12. A posthumous 1793 edition of James Thomson's *The Seasons* also included extensive annotations by Percival Stockdale, although these had less botanical specificity and were intended rather to provide cultural context for Thomson's mythological references. See Jung, *James Thomson's* The Seasons, 84–85.
13. Aikin, however, also dismissed botany as largely inappropriate for poetic study: "the vegetable creation, delightful as it is to the senses, and extensive in utility, yields comparatively few materials to the poet, whose art is principally defective in representing those qualities in which *it* chiefly excels; colour, scent, and taste" (*Essay on the Application*, 33).
14. Farina, "Vegetal Continuity," 421.
15. Farina, 421.
16. van der Veen, "Materiality of Plants," 808–9.
17. Marder, *Plant-Thinking*, 12.
18. Wakefield, *Introduction to Botany*, letter 2, 5.
19. Darwin, *Phytologia*, 136.

BIBLIOGRAPHY

Abercrombie, John. *Every Man His Own Gardener.* 16th ed. London: S. Hamilton, 1800.

———. *The Gardener's Pocket Dictionary.* London: n.p., 1786.

Adams, George. *Micrographia Illustrata.* London: Lockyer Davis, 1746.

Adams, Henry Gardiner. *A Cyclopædia of Female Biography: Consisting of Sketches of All Women Who Have Been Distinguished by Great Talents, Strength of Character, Piety, Benevolence, or Moral Virtue of Any Kind: Forming a Complete Record of Womanly Excellence or Ability.* London: Groombridge & Sons, 1857.

Aikin, John. *An Essay on the Application of Natural History to Poetry.* Warrington: W. Eyres, 1777.

———. *Manual of Materia Medica.* Yarmouth: n.p., 1785.

Alaimo, Stacy. *Undomesticated Ground: Recasting Nature as Feminist Space.* Ithaca, NY: Cornell University Press, 2000.

"Aloe vera." Plants of the World Online. Accessed May 21, 2021. http://powo.science.kew.org/taxon/urn:lsid:ipni.org:names:530017-1.

Armstrong, Nancy. "The Rise of the Domestic Woman." In *The Ideology of Conduct: Essays on Literature and the History of Sexuality,* edited by Nancy Armstrong and Leonard Tennenhouse, 96–141. New York: Methuen, 1987.

Bailes, Melissa. "Linnaeus's Botanical Clocks: Chronobiological Mechanisms in the Scientific Poetry of Erasmus Darwin, Charlotte Smith, and Felicia Hemans." *Studies in Romanticism* 56, no. 2 (2017): 223–52.

———. *Questioning Nature: British Women's Scientific Writing and Literary Originality, 1750–1830.* Charlottesville: University of Virginia Press, 2017.

Ballaster, Ros, Margaret Beetham, Elizabeth Frazer, and Sandra Hebron, eds. *Women's Worlds: Ideology, Femininity, and the Woman's Magazine.* London: Macmillan, 1991.

Baluška, Frantisek, Stefano Mancuso, and Dieter Volkmann. *Communication in Plants: Neuronal Aspects of Plant Life.* Berlin: Springer-Verlag Berlin Heidelberg, 2006.

Bannet, Eve Tavor. "Haywood's Spectator and the Female World." In Wright and Newman, eds., *Fair Philosopher,* 82–103.

Barbauld, Anna Laetitia. *Poems.* London: Joseph Johnson, 1773.

Baster, Job. "A Dissertation on the Worms which Destroy the Piles on the Coasts of Holland and Zealand." *Philosophical Transactions* 41, no. 455 (1740): 276–88.

Bataille, Robert R. *The Writing Life of Hugh Kelly*. Carbondale: Southern Illinois University Press, 2000.

Batchelor, Jennie, and Cora Kaplan. *Women and Material Culture, 1660–1830*. London: Palgrave Macmillan, 2007.

Behrendt, Stephen C. *British Women Poets and the Romantic Writing Community*. Baltimore, MD: Johns Hopkins University Press, 2018.

Beilby, Ralph. *History of British Birds*. Newcastle: Sol. Hodgson, 1797.

Belanger, Terry. "Publishers and Writers in Eighteenth-Century England." In *Books and Their Readers in Eighteenth-Century England*, edited by Isabel Rivers, 5–25. New York: St. Martin's, 1982.

Bending, Stephen. *Green Retreats: Women, Gardens, and Eighteenth-Century Culture*. Cambridge: Cambridge University Press, 2013.

Benjamin, Marina. "Elbow Room: Women Writers on Science, 1790–1840." In *Science and Sensibility: Gender and Scientific Enquiry 1780–1945*, edited by Marina Benjamin, 27–59. Oxford: Blackwell, 1991.

———. *A Question of Identity: Women, Science, and Literature*. New Brunswick, NJ: Rutgers University Press, 1993.

Bewell, Alan. "Erasmus Darwin's Cosmopolitan Nature." *ELH* 76, no. 1 (2009): 19–48.

———. "'Jacobin Plants': Botany as Social Theory in the 1790s." *Wordsworth Circle* 20, no. 3 (Summer 1989): 132–39.

———. *Natures in Translation*. Baltimore, MD: Johns Hopkins University Press, 2017.

———. "'On the Banks of the South Sea': Botany and Sexual Controversy in the Late Eighteenth Century." In *Visions of Empire: Voyages, Botany, and Representations of Nature*, edited by David Phillip Miller and Peter Manns Reill, 173–96. Cambridge: Cambridge University Press, 1996.

———. "Romanticism and Colonial Natural History." *Studies in Romanticism* 43, no. 1 (Spring 2004): 5–34.

Blacker, Mary Rose. *Flora Domestica*. London: National Trust Enterprises, 2000.

Blackwell, Alexander. *A New Method of Improving Cold, Wet, and Barren Lands: Particularly Clayey-Grounds*. London: n.p., 1741.

Blackwell, Elizabeth. *A Curious Herbal*. London: n.p., 1737. Copy at the University of Chicago Archives.

Blair, Patrick. *Pharmaco-Botanalogia*. London: G. Strahan, 1723–27.

Bleichmar, Daniela. *Visible Empire: Botanical Expeditions and Visual Culture in the Eighteenth-Century Hispanic Enlightenment*. Chicago: University of Chicago Press, 2012.

Blunt, Wilfred, and Sandra Raphael. *The Illustrated Herbal*. New York: Thames & Hudson, 1979.

Blunt, Wilfred, and William T. Stearn. *The Art of Botanical Illustration*. Woodbridge: Antique Collectors' Club, 1994.

Bonnet, Charles. "An Abstract of Some New Observations upon Insects: By M. Charles Bonnet of Geneva, Communicated in a Letter to Sir Hans Sloane, Bart. Late President of the Royal Society, &c. Translated from the French by P. H. Z. Esq; F. R. S." *Philosophical Transactions* 42, no. 470 (1743): 458–88.

Braam, Janet. "In Touch: Plant Responses to Mechanical Stimuli." *New Phytologist* 165, no. 2 (2004): 373–89.

Britten, James. "Biographical Notes LXVII: Mrs. Moriarty's 'Viridarium.'" *Journal of Botany, British and Foreign* 55 (1917): 52–54.

Brookes, Richard. *A New and Accurate System of Natural History*. London: J. Newbery, 1772.

Brown, Elizabeth Cullen. *The Sisters of St. Gothard*. London: Minerva, 1819.

Brunton, Mary. *Self-Control*. London: Longman, Hurst, Rees, Orme & Brown, 1811.

Butler, Marilyn. *Maria Edgeworth: A Literary Biography*. Oxford: Clarendon, 1972.

Carlile, Susan. *Charlotte Lennox: An Independent Mind*. Toronto: University of Toronto Press, 2018.

———. "Eyes That Eagerly 'Bear the Steady Ray of Reason': Eidolon as Activist in Charlotte Lennox's *The Lady's Museum*." In *Women's Periodicals and Print Culture in Britain, 1690–1820s*, edited by Jennie Batchelor and Manushag N. Powell, 357–76. Edinburgh: Edinburgh University Press, 2018.

A Catalogue of the Portland Museum. London: J. Bell, 1786.

Cavendish, Margaret. *The Blazing World and Other Writings*. New York: Penguin, 2004.

Chandler, James. "Edgeworth and the Lunar Enlightenment." *Eighteenth-Century Studies* 45, no.1 (2011): 87–104.

Chapman, Caroline. *Eighteenth-Century Women Artists*. London: Unicorn, 2017.

Chapone, Hester. *Letter to a New-Married Lady*. London: E. & C. Dilly, 1777.

Cheater, Christine. "Collectors of Nature's Curiosities: Science, Popular Culture, and the Rise of Natural History Museums." In *Frankenstein's Science: Experimentation and Discovery in Romantic Culture, 1780–1830*, edited by Christa Knellwolf and Jane Goddall, 167–82. Aldershot: Ashgate, 2008.

Cheston, Emily Reed. "Elizabeth Blackwell and Her 'Curious Herbal.'" *Herbarist: A Publication of the Herb Society of America* 8 (1942): 24–26.

Chico, Tita. *The Experimental Imagination*. Stanford, CA: Stanford University Press, 2018.

Clayton, Ellen Creathorne. *English Female Artists*. London: Tinsley Brothers, 1876.

Cohen, Jeffrey Jerome. "Monster Culture (Seven Theses)." In *Monster Theory: Reading Culture*, edited by Jeffrey Jerome Cohen, 3–25. Minneapolis: University of Minnesota Press, 1996.

A Companion to the Museum. London: Charles Elliot, 1790.

Cook, Elizabeth Heckendorn. "'Perfect' Flowers, Monstrous Women: Eighteenth-Century Botany and the Modern Gendered Subject." In *"Defects": Engendering the Modern Body*, edited by Helen Deutsch, 252–79. Ann Arbor: University of Michigan Press, 2000.

Coppola, Al. *The Theater of Experiment*. Oxford: Oxford University Press, 2016.

Cullen, William. *A Treatise of the Materia Medica*. Edinburgh: n.p., 1789.

Cunningham, Andrew. "The Culture of Gardens." In *Cultures of Natural History*, edited by N. Jardine, J. A. Secord, and E. C. Spary, 38–56. Cambridge: Cambridge University Press, 1996.

Curtis, William. *The Botanical Magazine; or, Flower-Garden Displayed*. Vol 10. London: Stephen Couchman, 1796.

———. *The Botanical Magazine; or, Flower-Garden Displayed*. Vol 14. London: Stephen Couchman, 1800.

Darwin, Erasmus. *The Botanic Garden*. London: n.p., 1790.

———. *Phytologia; or, The Philosophy of Agriculture and Gardening*. London: J. Johnson, 1800.

Daston, Lorraine, and Elizabeth Lunbeck, eds. *Histories of Scientific Observation*. Chicago: University of Chicago Press, 2011.

Daston, Lorraine, and Peter Galison. *Objectivity*. New York: Zone, 2010.

Davis, Natalie Zemon. *Women on the Margins*. Cambridge, MA: Harvard University Press, 1995.

Delany, Mary. *The Autobiography and Correspondence of Mary Granville, Mrs. Delany*. Edited by Lady Augusta Llanover. 6 vols. London: Richard Bentley, 1861–62.

Delaporte, Francis. *Nature's Second Kingdom: Explorations of Vegetality in the Eighteenth Century*. Translated by Arthur Goldhammer. Cambridge, MA: MIT Press, 1982.

Dorn, Judith. "Reading Women Reading History: The Philosophy of Periodical Form in Charlotte Lennox's *The Lady's Museum*." *Historical Reflections* 18, no. 3 (1992): 7–27.

Douthwaite, Julia. "Experimental Child-Rearing after Rousseau: Maria Edgeworth, Practical Education, and *Belinda*." *Irish Journal of Feminist Studies* 2, no. 2 (Winter 1997): 35–56.

———. *The Wild Girl, Natural Man, and the Monster: Dangerous Experiments in the Age of Enlightenment*. Chicago: University of Chicago Press, 2002.

Dudley, Susan A., and Amanda L. File. "Kin Recognition in an Annual Plant." *Biology Letters* 3, no. 4 (2007): 435–48.

Edmonson, John. "Novelty in Nomenclature: The Botanical Horizons of Mary Delany." In Laird and Weisberg-Roberts, eds., *Mrs. Delany and Her Circle*, 188–203.

Edgeworth, Maria. *Belinda*. Edited by Kathryn J. Kirkpatrick. Oxford: Oxford University Press, 2008.

———. "The Good French Governess," In *Tales and Novels, Vol. 1: Moral Tales.* Hildesheim: G. Olms, 1969.

———. *Letters for Literary Ladies.* London: Everyman, 1993.

———. *Maria Edgeworth's Letters from Ireland.* Edited by Valerie Packenham. Dublin: Lilliput, 2018.

———. *The Parent's Assistant; Moral Tales for Young People.* In *The Novels and Selected Works of Maria Edgeworth.* Vol. 10. Edited by Elizabeth Eger and Clíona Ó Gallchoir. London: Pickering & Chatto, 2003.

———. *Practical Education.* In *The Novels and Selected Works of Maria Edgeworth.* Vol. 11. Edited by Susan Manly. London: Pickering & Chatto, 2003.

Fara, Patricia. *Pandora's Breeches: Women, Science, and Power in the Enlightenment.* London: Pimlico, 2004.

———. *Sex, Botany, and Empire: The Story of Carl Linnaeus and Joseph Banks.* New York: Columbia University Press, 2004.

Farina, Lara. "Vegetal Continuity and the Naming of the Species." *Postmedieval: A Journal of Medieval Cultural Studies* 9, no. 4 (2018): 420–31.

Fennetaux, Ariane. "Female Crafts: Women and Bricolage in Late Georgian Britain, 1750–1820." In *Women and Things,* edited by Maureen Daly Goggin and Beth Fowkes Tobin, 91–108. Aldershot: Ashgate, 2009.

Fergus, Jan. "Women Readers: A Case Study." In *Women and Literature in Britain, 1700–1800,* edited by Vivien Jones, 155–76. Cambridge: Cambridge University Press, 2000.

Findlen, Paula. *Possessing Nature: Museums, Collecting, and Scientific Culture in Early Modern Italy.* Berkeley: University of California Press, 1994.

Ford, Lisa. "A Progress in Plants." In Laird and Weisberg-Roberts, eds., *Mrs. Delany and Her Circle,* 204–23.

Foucault, Michel. *The Order of Things: An Archaeology of the Human Sciences.* New York: Vintage, 1994.

Gagliano, Monica. *The Language of Plants.* Minneapolis: University of Minnesota Press, 2017.

Gates, Barbara T. *Kindred Nature: Victorian and Edwardian Women Embrace the Living World.* Chicago: University of Chicago Press, 1998.

Gates, Barbara T., and Ann B. Shteir. *Natural Eloquence: Women Reinscribe Science.* Madison: University of Wisconsin Press, 1997.

Gentleman of the Faculty. *Every Lady Her Own Physician; or, The Closet Companion.* London: M. Randall, 1788.

A Genuine Copy of a Letter from a Merchant in Stockholm, to His Correspondent in London, Containing an Impartial Account of Doctor Alexander Blackwell, His Plot, Trial, Character, and Behaviour, Both under Examination and at the Place of Execution. London: H. Carpenter, 1747.

George, Sam. *Botany, Sexuality, and Women's Writing, 1760–1830.* Manchester: Manchester University Press, 2007.

———. "Carl Linnaeus, Erasmus Darwin, and Anna Seward: Botanical Poetry and Female Education." *Science & Education* 23, no. 3 (2014): 673–94.

Gere, Charlotte, and Marina Vaizey. *Great Women Collectors.* New York: Harry N. Abrams, 1999.

Gevirtz, Karen. *Women, the Novel, and Natural Philosophy, 1660–1727.* New York: Palgrave Macmillan, 2014.

Gibson, Prudence. *The Plant Contract.* Leiden: Brill, 2018.

Gidal, Eric. *Ossianic Unconformities: Bardic Poetry in the Industrial Age.* Charlottesville: University of Virginia Press, 2015.

Gilmore, John. *The Poetics of Empire: A Study of James Grainger's* The Sugar Cane *(1764).* London: Athalone, 2000.

Girten, Kristen M. "Unsexed Souls: Natural Philosophy as Transformation in Eliza Haywood's *Female Spectator.*" *Eighteenth-Century Studies* 43, no. 1 (Fall 2009): 55–74.

Goggin, Maureen Daly, and Beth Fowkes Tobin. *Women and the Material Culture of Needlework and Textiles, 1750–1950.* Farnham: Ashgate, 2009.

———. *Women and Things, 1750–1950: Gendered Material Strategies.* Farnham: Ashgate, 2009.

Gonda, Caroline. *Reading Daughter's Fictions, 1709–1834: Novels and Society from Manley to Edgeworth.* Cambridge: Cambridge University Press, 1996.

Goodman, Kevis. "Conjectures on *Beachy Head:* Charlotte Smith's Geological Poetics and the Ground of the Present." *ELH* 81, no. 3 (Fall 2014): 983–1006.

Gordon, Scott Paul. *The Practice of Quixotism.* New York: Palgrave Macmillan, 2006.

———. "The Space of Romance in Lennox's *Female Quixote.*" *SEL* 38, no. 3 (1998): 499–516.

Gosse, Ellen. "Of Paper Flowers." *Temple Bar: A London Magazine for Town and Country Readers* 112, no. 441 (December 1897): 506–18.

Gowrley, Freya. "Craft(ing) Narratives: Specimens, Souvenirs, and 'Morsels' in A la Ronde's Specimen Table." *Eighteenth-Century Fiction* 31, no. 1 (Fall 2018): 77–97.

Greenfield, Susan C. "'Abroad and at Home': Sexual Ambiguity, Miscegenation, and Colonial Boundaries in Edgeworth's *Belinda.*" *PMLA* 112, no. 2 (1997): 214–48.

Hales, Stephen. *Vegetable Staticks.* London: W. & J. Innys, 1727.

Hall, Marie Boas. *Promoting Experimental Learning: Experiment and the Royal Society, 1660–1727.* Cambridge: Cambridge University Press, 1991.

Hall, Matthew. *Plants as Persons: A Philosophical Botany.* Albany: State University of New York Press, 2011.

Halsey, Katie. "The Home Education of Girls in the Eighteenth-Century Novel: The Pernicious Effects of an Improper Education." *Oxford Review of Education* 41, no. 4 (2015): 430–46.

Haraway, Donna. *Primate Visions: Gender, Race, and Nature in the World of Modern Science.* New York: Routledge, 1989.

———. *Simians, Cyborgs, and Women: The Reinvention of Nature.* New York: Routledge, 1990.

Harding, Sandra. *The Science Question in Feminism.* Ithaca, NY: Cornell University Press, 1986.

Harris, Stephen. *Planting Paradise: Cultivating the Garden, 1501–1900.* Oxford: Bodleian Library, 2011.

Haut, Asia. "Reading Flora: Erasmus Darwin's The Botanic Garden, Henry Fuseli's Illustrations, and Various Literary Responses." *Word & Image: A Journal of Verbal/Visual Enquiry* 20, no. 4 (2012): 240–56.

Hayden, Judy A. ed. *The New Science and Women's Literary Discourse: Prefiguring* Frankenstein. New York: Palgrave Macmillan, 2011.

Hayden, Ruth. *Mrs. Delany: Her Life and Her Flowers.* New York: New Amsterdam, 1980.

Haywood, Eliza. *The Female Spectator.* In *Selected Works of Eliza Haywood,* series 2, vols. 2–3. Edited by Kathryn R. King and Alexander Pettit. London: Pickering & Chatto, 2001.

———. *Selections from the* Female Spectator. Edited by Patricia Meyer Spacks. Oxford: Oxford University Press, 1999.

Heringman, Noah. *Romantic Rocks, Aesthetic Geology.* Ithaca, NY: Cornell University Press, 2010.

Heyd, Thomas, ed. *Recognizing the Autonomy of Nature: Theory and Practice.* New York: Columbia University Press, 2005.

Hilton, Mary. *Women and the Shaping of the Nation's Young: Education and Public Doctrine in Britain, 1750–1850.* New York: Routledge, 2016.

Hollis, Karen. "Eliza Haywood and the Gender of Print." *Eighteenth Century* 38 (1997): 43–62.

Holmes, Richard. *The Age of Wonder: How the Romantic Generation Discovered the Beauty and Terror of Science.* New York: Pantheon, 2009.

Houle, Karen. "Animal, Vegetable, Mineral: Ethics as Extension or Becoming? The Case of Becoming-Plant." *Journal for Critical Animal Studies* 9 nos. 1–2 (2011): 89–116.

Italia, Iona. *The Rise of Literary Journalism in the Eighteenth Century: Anxious Employment.* New York: Routledge, 2005.

Jackson, Christine E. *Sarah Stone: Natural Curiosities from the New Worlds.* London: Merrell Holberton/Natural History Museum, London, 1998.

Jacson, Maria. *Botanical Dialogues.* London: J. Johnson, 1797. Copy at McMaster University Archives.

Jouslin, Claire Boulard. "Conservative or Reformer? The History and Fortune of Fénelon's Traite de l'Education des Filles in Eighteenth-Century England." *Journal of Early Modern Cultural Studies* 12, no. 4 (2012): 48–77.

Jung, Sandro. *James Thomson's The Seasons, Print Culture, and Visual Interpretation, 1730–1842*. Bethlehem, PA: Lehigh University Press, 2015.

Kaul, Suvir. "Why Selima Drowns: Thomas Gray and the Domestication of the Imperial Ideal." *PMLA* 102, no. 2 (1990): 223–32.

Keeler, Nancy Boghossian. *Gardens in Perpetual Bloom: Botanical Illustration in Europe and America, 1600–1850*. Boston: MFA, 2010.

Keen, Paul. *The Crisis of Literature in the 1790s: Print Culture and the Public Sphere*. Cambridge: Cambridge University Press, 1999.

Keetley, Dawn. "Introduction: Six Theses on Plant Horror." In *Plant Horror: Approaches to the Monstrous Vegetal in Fiction and Film*, edited by Dawn Keetley and Angela Tenga, 1–30. London: Palgrave Macmillan, 2016.

Keller, Evelyn Fox. *Reflections on Gender and Science*. New Haven, CT: Yale University Press, 1985.

———. *Secrets of Life/Secrets of Death: Essays on Language, Gender, and Science*. New York: Routledge, 1992.

Kelley, Theresa M. *Clandestine Marriage: Botany and Romantic Culture*. Baltimore, MD: Johns Hopkins University Press, 2012

———. "Romantic Exemplarity: Botany and 'Material' Culture." In *Romantic Science: The Literary Forms of Natural History*, edited by Noah Heringman, 223–54. Albany: State University of New York Press, 2003.

King, Amy M. *Bloom: The Botanical Vernacular in the English Novel*. Oxford: Oxford University Press, 2003.

King, Kathryn R. *A Political Biography of Eliza Haywood*. London: Pickering & Chatto, 2012.

King, Ronald. *Botanical Illustration*. New York: Clarkson N. Potter, 1979.

Kippis, Andrew. "Delany, Mary." In *Biographia Britannica: or, The Lives of the Most Eminent Persons Who Have Flourished in Great-Britain and Ireland from the Earliest Ages, to the Present Times*, 88–93. Vol. 2. 2nd ed. London: John Nichols, 1793.

Kirkpatrick, Kathryn J. "The Limits of Liberal Feminism in Maria Edgeworth's *Belinda*." In *Jane Austen and Mary Shelley and Their Sisters*, edited by Laura Dabundo, 73–82. New York: University Press of America, 2000.

Klein, Ula. "Bosom Friends and the Sapphic Breasts of *Belinda*." *ABO* 3, no. 2 (2013): https://scholarcommons.usf.edu/abo/vol3/iss2/1/ (accessed June 14, 2021).

Knowles, Claire, and Ingrid Horricks, eds. *Charlotte Smith: Major Poetic Works*. Peterborough, ON: Broadview, 2017.

Koerner, Lisbet. "Carl Linnaeus in His Time and Place." In *Cultures of Natural History*, edited by N. Jardine, J. A. Secord, and E. M. Spary, 145–62. Cambridge: Cambridge University Press, 1996.

———. *Linnaeus: Nature and Nation*. Cambridge, MA: Harvard University Press, 1999.

Kohn, Eduardo. *How Forests Think: Toward an Anthropology beyond the Human.* Oakland: University of California Press, 2013.

Kowaleski-Wallace, Elizabeth. *Their Fathers' Daughters: Hannah More, Maria Edgeworth, and Patriarchal Complicity.* Oxford: Oxford University Press, 1991.

Kramer, Jack. *Women of Flowers: A Tribute to Victorian Women Illustrators.* New York: Stewart, Tabori & Chang, 1996.

Kranz, Isabel. "The Language of Flowers in Popular Culture and Botany." In *The Language of Plants: Science, Philosophy, and Literature,* edited by Monica Gagliano, John C. Ryan, and Patricia Vieira, 193–214. Minneapolis: University of Minnesota Press, 2017.

La Mettrie, Julian Offray de. *Machine Man and Other Writings.* Edited and translated by Anna Thomson. New York: Cambridge University Press, 1996.

Labbe, Jacqueline M. *Charlotte Smith: Romanticism, Poetry, and the Culture of Gender.* Manchester: Manchester University Press, 2003.

———. *Romantic Visualities: Landscape, Gender, and Romanticism.* Manchester: Manchester University Press, 2003.

Laird, Mark. *A Natural History of English Gardening, 1650–1800.* New Haven, CT: Yale University Press, 2015.

Laird, Mark, and Alicia Weisberg-Roberts, eds. *Mrs. Delany and Her Circle.* New Haven, CT: Yale Center for British Art, 2009.

Langbauer, Laurie. *Women and Romance: The Consolations of Gender in the English Novel.* Ithaca, NY: Cornell University Press, 1990.

Lawrence, Anne. *Women in England, 1500–1760: A Social History.* New York: St. Martin's, 1994.

Lee, James. *An Introduction to Botany.* London: J. & R. Tonson, 1760. Copy at McMaster University Archives.

"Lemna." Plants of the World Online. Accessed May 18, 2021. http://powo.science.kew.org/taxon/urn:lsid:ipni.org:names:23865-1.

Lennox, Charlotte. *Charlotte Lennox: Correspondence and Miscellaneous Documents.* Edited by Norbert Schürer. Lewisburg, PA: Bucknell University Press, 2012.

———. *The Female Quixote.* Oxford: Oxford University Press, 2008.

———. *Shakespear Illustrated.* London: A. Millar, 1753.

———. *Sophia.* Edited by Norbert Schürer. Peterborough, ON: Broadview, 2008.

Leong, Elaine. *Recipes and Everyday Knowledge: Medicine, Science, and the Household in Early Modern England.* Chicago: University of Chicago Press, 2018.

Leong, Elaine, and Alisha Rankin, eds. *Secrets and Knowledge in Medicine and Science, 1500–1800.* New York: Routledge, 2011.

Linkin, Harriet Kramer, and Stephen C. Behrendt, eds. *Romanticism and Women Poets: Opening the Doors of Reception.* Lexington: University of Kentucky Press, 1999.

Linnaean Collections. Accessed May 20, 2021. https://www.linnean.org/research-collections/linnaean-collections.

Linnaeus. *Species Plantarum*. Holmiae: Impensis Laurentii Salvii, 1753.

Looser, Devoney. *British Women Writers and the Writing of History, 1670–1820*. Baltimore, MD: Johns Hopkins University Press, 2000.

Lorenzo-Modia, María Jesús. "Education for Women in the Eighteenth-Century Periodical: Charlotte Lennox's *The Lady's Museum*." In *Diferencia, (des)igualdad y justiciar/Differences, (in)equality, and justice*, edited by A. Antón Pacheco, I. Durán Giménez-Rico, C. Méndez García, J. Neff van Aertselaer, and A. L. Rodríguez Redondo, 201–12. Madrid: Editorial Fundamentos, 2010.

Lynch, Deidre Shauna. "'Young Ladies Are Delicate Plants': Jane Austen and Greenhouse Romanticism." *ELH* 77 no. 3 (Fall 2010): 689–729.

Mabey, Richard. *The Cabaret of Plants: Forty Thousand Years of Plant Life and the Human Imagination*. New York: W. W. Norton, 2017.

MacDonald, Edgar E. *The Education of the Heart: The Correspondence of Rachel Mordecai Lazarus and Maria Edgeworth*. Chapel Hill: University of North Carolina Press, 1977.

MacFayden, Heather. "Lady Delacour's Library: Maria Edgeworth's *Belinda* and Fashionable Reading." *Nineteenth-Century Literature* 48, no. 4 (March 1994): 423–39.

Mack, Ruth. "Quixotic Ethnography: Charlotte Lennox and the Dilemma of Cultural Observation." *Novel: A Forum on Fiction* 38, nos. 2–3 (2005): 193–213.

Madge, Bruce. "Elizabeth Blackwell—the Forgotten Herbalist?" *Health Information and Libraries Journal* 18, no. 3 (2003): 144–52.

Mahood, M. M. *The Poet as Botanist*. Cambridge: Cambridge University Press, 2008.

Mancuso, Stefano. *The Revolutionary Genius of Plants: A New Understanding of Plant Intelligence and Behavior*. New York: Atria, 2018.

Mancuso, Stefano, and Alessandra Viola. *Brilliant Green: The Surprising History and Science of Plant Intelligence*. London: Island, 2005.

Marder, Michael. *Plant-Thinking: A Philosophy of Vegetal Life*. New York: Columbia University Press, 2013.

Marder, Michael, and Luce Irigaray. *Through Vegetal Being*. New York: Columbia University Press, 2016.

Mathias, Thomas James. *The Pursuits of Literature: A Satirical Poem in Four Dialogues*. London: T. Becket, 1794–97.

Maurer, Shawn Lisa. *Proposing Men: Dialectics of Gender and Class in the Eighteenth-Century England Periodical*. Stanford, CA: Stanford University Press, 1998.

Mellor, Anne K. *Romanticism and Gender*. New York: Routledge, 1993.

Merchant, Carolyn. *The Death of Nature: Women, Ecology, and the Scientific Revolution*. New York: HarperCollins, 1980.

Miguel-Alfonso, Ricardo. "Social Conservatism, Aesthetic Education, and the Essay Genre in Eliza Haywood's *The Female Spectator*." In Wright and Newman, eds., *Fair Philosopher*, 72–81.

Miller, Elaine. *The Vegetative Soul: From Philosophy of Nature to Subjectivity in the Feminine*. Albany: State University of New York Press, 2002.

Miller, Philip. *The Gardeners Dictionary*. 6th ed. London: n.p., 1752.

Miller, T. S. "Lives of the Monster Plants: The Revenge of the Vegetable in the Age of Animal Studies." *Journal of the Fantastic in the Arts* 23, no. 3 (2012): 460–79.

Montwieler, Katherine. "Reading Disease: The Corrupting Performance of Edgeworth's *Belinda*." *Women's Writing* 12, no. 3 (2005): 347–68

Moore, Lisa. "Queer Gardens: Mary Delany's Flowers and Friendships." *Eighteenth-Century Studies* 39, no. 1 (Fall 2005): 49–70.

———. *Sister Arts: The Erotics of Lesbian Landscapes*. Minneapolis: University of Minnesota Press, 2011.

Moriarty, Henrietta Maria. *Brighton in an Uproar*. London: W. Glindon, 1811. Copy at the University of Illinois Rare Book and Manuscript Library.

———. *Fifty Plates of Green-House Plants*. London: T. Bensley, 1807.

Murray, Lady Charlotte. *The British Garden*. London: S. Hazard, 1799. Copy at McMaster University Archives.

Myers, Mitzi. "Impeccable Governesses, Rational Dames, and Moral Mothers: Mary Wollstonecraft and the Female Tradition in Georgian Children's Books." *Children's Literature* 14 (1986): 31–59.

Narain, Mona. "Not the Angel in the House: Intersections of Public and Private in Maria Edgeworth's *Moral Tales* and *Practical Education*." In *New Essays on Maria Edgeworth*, edited by Julie Nash, 57–71. Aldershot: Ashgate, 2006.

———. "A Prescription of Letters: Maria Edgeworth's *Letters for Literary Ladies* and the Ideologies of the Public Sphere." *Journal of Narrative Technique* 28, no. 3 (Fall 1998): 266–86.

Nealon, Jeffrey. *Plant Theory: Biopower and Vegetable Life*. Stanford, CA: Stanford University Press, 2016.

A New Treatise on Flower Painting: Containing the Most Familiar and Easy Instructions; with Directions How to Mix the Various Tints, Etc. London: n.p., 1797.

Nollet, Jean-Antoine. *Leçons de physique expérimentale*. Paris: n.p., 1743.

———. "Part of a Letter from Abbe Nollet, of the Royal Academy of Sciences at Paris, and F. R. S. to Martin Folkes Esq; Presisident [*sic*] of the Same, concerning Electricity," *Philosophical Transactions of the Royal Society* 45, no. 486 (1748): 187–94.

Nyberg, Kenneth. "Linnaeus's Apostles and the Globalization of Knowledge, 1729–1756." In *Global Scientific Practice in an Age of Revolutions, 1750–1850*, edited by Patrick Manning and Daniel Rood, 73–89. Pittsburgh: University of Pittsburgh Press, 2016.

Ó Gallchoir, Clíona. "Gender, Nation, and Revolution: Maria Edgeworth and Stéphanie-Félicité de Genlis." In *Women, Writing, and the Public Sphere, 1700–1830*, edited by Elizabeth Eger, Charlotte Grant, Clíona Ó Gallchoir, and Penny Warburton, 200–216. Cambridge: Cambridge University Press, 2001.

O'Malley, Therese, and Amy R. W. Meyers. *The Art of Natural History: Illustrated Treatises and Botanical Paintings, 1400–1850*. New Haven, CT: Yale University Press, 2008.

Ogilvie, Brian. *The Science of Describing*. Chicago: University of Chicago Press, 2006.

Orr, Clarissa Campbell. *Mrs. Delany: A Life*. New Haven, CT: Yale University Press, 2019.

Page, Judith W., and Elise L. Smith, eds. *Women, Literature, and the Domesticated Landscape: England's Disciples of Flora, 1780–1870*. Cambridge: Cambridge University Press, 2011.

Palo, Sharon Smith. "The Good Effects of a Whimsical Study: Romance and Women's Learning in Charlotte Lennox's *The Female Quixote*." *ECF* 18 no. 2 (2005–6): 203–28.

Pascoe, Judith. "Female Botanists and the Poetry of Charlotte Smith." In *Re-Visioning Romanticism: British Women Writers, 1776–1837*, edited by Carol Shiner Wilson and Joel Haefner, 193–209. Philadelphia: University of Pennsylvania Press, 1994.

Peacock, Molly. *The Paper Garden: Mrs. Delany [Begins Her Life's Work] at 72*. New York: Bloomsbury, 2011.

Pearson, Jacqueline. *Women's Reading in Britain, 1750–1835*. Cambridge: Cambridge University Press, 1999.

Pelling, Madeleine. "Collecting the World: Female Friendship and Domestic Craft at Bulstrode Park." *Journal for Eighteenth-Century Studies* 41, no. 1 (2018): 1–20.

———. *The Duchess' Museum: Collecting, Craft and Conversation*. Forthcoming.

Phillips, Patricia. *The Scientific Lady: A Social History of Women's Scientific Interests, 1520–1918*. London: Weidenfeld & Nicolson, 1990.

Plant: Exploring the Botanical World. London: Phaidon, 2016.

Polwhele, Richard. *The Unsex'd Females; a Poem, Addressed to the Author of* The Pursuits of Literature. New York: Wm. Cobbett, 1800.

Pope, Alexander. *The Major Works*. Edited by Pat Rogers. Oxford: Oxford University Press, 2008.

Prescott, Sarah, and Jane Spencer. "Prattling, Tattling, and Knowing Everything: Public Authority and the Female Editorial Persona in the Early Essay-Periodical." *Journal for Eighteenth-Century Studies* 23, no. 1 (2000): 43–57.

"Prunus cerasus." Plants of the World Online. May 21, 2021. http://powo.science.kew.org/taxon/urn:lsid:ipni.org:names:729574-1.

Pulteney, Richard. *Historical and Biographical Sketches of the Progress of Botany in England, from Its Origin to the Introduction of the Linnæan System*. London: T. Cadell, 1790.

Purvor, Margery. *The Royal Society: Concept and Creation*. London: Routledge, 1967.

Pyle, Hilary. "Artist or Artistic? The Drawings of Mary Delany." *Irish Arts Review* (1987): 27–32.

Raven, James, Helen Small, and Naomi Tadmor, eds. *The Practice and Representation of Reading*. Cambridge: Cambridge University Press, 1996.

Ray, John. *Historia Plantarum*. London: n.p., 1683.

Reeder, Kohleen. "The 'Paper Mosaick' Practice of Mrs. Delany's Circle." In Laird and Weisberg-Roberts, eds., *Mrs. Delany and Her Circle*, 225–35.

Reeve, Clara. *The Progress of Romance*. Colchester: n.p., 1785.

"A Register of Books Published in July." *Gentleman's Magazine* 7 (July 1737): 454.

Roberts, Felicity. "'Idleness Never Grew in My Soil': Mary Delany's Flower Collages, Gender, and the Moral Authority of 'Nature' in Eighteenth-Century England." In *Materializing Gender in Eighteenth-Century Europe*, edited by Jennifer G. Germann and Heidi Strobel, 141–56. New York: Routledge, 2016.

Rogers, Katherine. *Feminism in Eighteenth-Century England*. Urbana: University of Illinois Press, 1982.

Ross, Stephanie. *What Gardens Mean*. Chicago: University of Chicago Press, 1998.

Rousseau, G. S. "Science Books and Their Readers in the Eighteenth Century." In *Books and Their Readers in Eighteenth-Century England*, edited by Isabel Rivers, 197–255. New York: St. Martin's, 1982.

Rousseau, Jean-Jacques. *Letters on the Elements of Botany*. Translated by Thomas Martyn. London: B. White & Son, 1794.

Rowden, Frances Arabella. *Poetical Introduction to the Study of Botany*. London: T. Bensley, 1801. Copy at the Newberry Library.

Ruwe, Donelle R. "Charlotte Smith's Sublime: Feminine Poetics, Botany, and *Beachy Head*." *Essays in Romanticism* 7, no. 1 (1999): 117–32.

Sagal, Anna K. "Constructing Women's History in *The Lady's Museum*." In *Women's Periodicals and Print Culture in Britain, 1690–1820s: The Long Eighteenth Century*, edited by Jennie Batchelor and Manushag N. Powell, 53–66. Edinburgh: Edinburgh University Press, 2018.

Saunders, Gil. *Picturing Plants: An Analytical History of Botanical Illustration*. London: KWS, 2009.

Scantlebury, Kathryn, and Collette Murphy. "Maria Edgeworth: Nineteenth-Century Irish Female Pioneer of Science Education." *Irish Educational Studies* 28, no. 1 (March 2009): 103–13.

Schiebinger, Londa. *The Mind Has No Sex? Women in the Origins of Modern Science*. Cambridge, MA: Harvard University Press, 1992.

———. *Nature's Body: Gender in the Making of Modern Science*. Boston: Beacon, 1993.

———. *Plants and Empire: Colonial Bioprospecting in the Atlantic World.* Cambridge, MA: Harvard University Press, 2004.

———. "The Private Life of Plants: Sexual Politics in Carl Linnaeus and Erasmus Darwin." In *Science and Sensibility: Gender and Scientific Enquiry 1780–1945*, edited by Marina Benjamin, 121–43. Oxford: Basil Blackwell, 1991.

Scott, Sarah. *Millenium Hall.* Edited by Gary Kelly. Peterborough, ON: Broadview, 2004.

Shapin, Stephen, and Simon Schaffer. *Leviathan and the Air-Pump: Hobbes, Boyle, and the Experimental Life.* Princeton, NJ: Princeton University Press, 1985.

Shaw, George, and Frederick P. Nodder, *The Naturalist's Miscellany, or Coloured Figures of Natural Objects.* London: Nodder, 1789–1813.

Shevelow, Kathryn. "'C—L—' to 'Mrs. Stanhope': A Preview of Charlotte Lennox's *The Lady's Museum.*" *Tulsa Studies in Women's Literature* 1, no. 1 (1982): 83–86.

———. *Women and Print Culture: The Construction of Femininity in the Early Periodical.* New York: Routledge, 1990.

Shirk, Henrietta Nickels. "Contributions to Botany, the Female Science, by Two Eighteenth-Century Women Technical Communicators." *Technical Communications Quarterly* 6 no. 3 (1997): 293–312.

Shteir, Ann B. "Botanical Dialogues: Maria Jacson and Women's Popular Science Writing in England." *Eighteenth-Century Studies* 23, no. 3 (1990): 301–17.

———. *Cultivating Women, Cultivating Science: Flora's Daughters and Botany in England, 1760–1860.* Baltimore, MD: Johns Hopkins University Press, 1996.

Sloan, Kim. *"A Noble Art": Amateur Artists and Drawing Masters, c. 1600–1800.* London: British Museum, 2000.

Sloboda, Stacey. "Displaying Materials: Porcelain and Natural History in the Duchess of Portland's Museum." *Eighteenth-Century Studies* 43, no. 4 (Summer 2010): 455–72.

Smith, Charlotte. *The Collected Letters of Charlotte Smith.* Edited by Judith Philips Stanton. Bloomington: Indiana University Press, 2004.

———. "Flora." In Knowles and Horrocks, eds., *Charlotte Smith*, 203–10. Peterborough, ON: Broadview, 2017.

———. *Marchmont.* London: n.p., 1796.

Smith, James Edward. *Icones pictae plantarum rariorum descriptionibus et observationibus illustratae.* London: n.p., 1790–93.

———. *Lachesis Lapponica, or a Tour in Lapland, Now First Published from the Original Manuscript Journal of the Celebrated Linnaeus.* London: White and Cochrane, 1811.

Smith, Orianne. *Romantic Women Writers, Revolution, and Prophecy: Rebellious Daughters, 1786–1826.* Cambridge: Cambridge University Press, 2013.

Smith, Pamela H., ed. *Entangled Itineraries: Materials, Practices, and Knowledges across Eurasia.* Pittsburgh: University of Pittsburgh Press, 2019.

———. "Making as Knowing: Craft as Natural Philosophy." In Smith, Meyers, and Cook, eds., *Ways of Making and Knowing*, 17–47.

Smith, Pamela H., Amy R. W. Meyers, and Harold J. Cook, eds. *Ways of Making and Knowing: The Material Culture of Empirical Knowledge*. Ann Arbor: University of Michigan Press, 2014.

The Snail; or, The Lady's Lucubrations. London: E. Boyd, 1745.

Snook, Edith. "'The Beautifying Part of Physic': Women's Cosmetic Practices in Early Modern England." *Journal of Women's History* 20, no. 3 (Fall 2008): 10–33.

Spacks, Patricia Meyer. *Selections from the* Female Spectator. Oxford: Oxford University Press, 1999.

Spedding, Patrick. "Measuring the Success of the *Female Spectator*." In Wright and Newman, eds., *Fair Philosopher*, 193–211.

Stafford, William. *English Feminists and Their Opponents in the 1790s*. Manchester: Manchester University Press, 2002.

Sweet, Nanora, and Julie Melnyk. *Felicia Hemans: Reimagining Poetry in the Nineteenth Century*. New York: Palgrave Macmillan, 2001.

Swift, Jonathan. *The Lady's Dressing Room*. 2nd ed. Dublin: J. Roberts, 1732.

Tayebi, Kandi. "Undermining the Eighteenth-Century Pastoral: Rewriting the Poet's Relationship to Nature in Charlotte Smith's Poetry." *European Romantic Review* 15, no. 1 (March 2004): 131–50.

Terrall, Mary. "Following Insects Around: Tools and Techniques of Eighteenth-Century Natural History." *British Journal for the History of Science* 43, no. 4 (December 2010): 573–88.

Teute, Fredrika J. "The Loves of the Plants: or, The Cross-Fertilization of Science and Desire in the Eighteenth Century." *Huntington Library Quarterly* 63, no. 3 (2000): 319–45.

Thompson, Helen. *Fictional Matter: Empiricism, Corpuscles, and the Novel*. Philadelphia: University of Pennsylvania Press, 2017.

Toal, Catherine. "Control Experiment: Edgeworth's Critique of Rousseau's Educational Theory." In *An Uncomfortable Authority: Maria Edgeworth and Her Contexts*, edited by Heidi Kaufman and Chris Fauske, 212–31. Newark: University of Delaware Press, 2004.

Tobin, Beth Fowkes. *The Duchess's Shells: Natural History Collecting in the Age of Cook's Voyages*. New Haven, CT: Paul Mellon Center, 2013.

———. "Imperial Designs: Botanical Illustration and the British Botanic Empire." *Studies in Eighteenth-Century Culture* 25 (1996): 265–92.

Trew, Christoph Jacob, George Dionysius Ehret, and Johann Jacob Haid. *Plantae selectae quarum imagines ad exemplaria naturalia Londini, in hortis curiosorum nutrita*. n.p.: n.p., 1750–73.

Turner, Gerard L'Estrange. *Scientific Instruments, 1500–1900: An Introduction*. Berkeley: University of California Press, 1998.

Uglow, Jennifer S., *The Lunar Men: Five Friends Whose Curiosity Changed the World*. New York: Farrar, Straus & Giroux, 2002.

———. *Nature's Engraver: A Life of Thomas Bewick*. London: Faber & Faber, 2006.

Valenza, Robin. *Literature, Language, and the Rise of the Intellectual Disciplines in Britain, 1680–1820*. Cambridge: Cambridge University Press, 2009.

van der Veen, Marijke. "The Materiality of Plants: Plant-People Entanglements." *World Archaeology* 46, no. 5 (2014): 799–812.

Vickery, Amanda. *The Gentleman's Daughter*. New Haven, CT: Yale University Press, 1998.

———. "The Theory and Practice of Female Accomplishment." In Laird and Weisberg-Roberts, eds., *Mrs. Delany and Her Circle*, 94–109.

Wakefield, Priscilla. *An Introduction to Botany*. London, 1796. Copy at McMaster University Archives.

Wall, Wendy. *Recipes for Thought: Knowledge and Taste in the Early Modern English Kitchen*. Philadelphia: University of Pennsylvania Press, 2015.

Watts, Ruth, "Whose Knowledge? Gender, Education, Science, and History." *History of Education* 36, no. 3 (May 2007): 283–302.

Weisberg-Roberts, Alicia. "Between Trade and Science: Dyeing and Knowing in the Long Eighteenth Century." In Smith, Meyers, and Cook, eds., *Ways of Making and Knowing*, 86–112.

Weiss, Deborah. "The Extraordinary Ordinary Belinda: Maria Edgeworth's Female Philosopher." *Eighteenth-Century Fiction* 19, no. 4 (Summer 2007): 441–61.

Weiss-Smith, Courtney. *Empiricist Devotions: Science, Religion, and Poetry in Early Eighteenth-Century England*. Charlottesville: University of Virginia Press, 2016.

Wirt, Elizabeth Washington. *Flora's Dictionary*. Baltimore: Fielding Lucas Jr., 1837.

Wollstonecraft, Mary. *A Vindication of the Rights of Woman*. Edited by Janet Todd. Oxford: Oxford University Press, 2008.

Wright, Lynn Marie, and Donald J. Newman, eds. *Fair Philosopher: Eliza Haywood and* The Female Spectator. Lewisburg: Bucknell University Press, 2006.

Wright, Nicole M. "Opening the Phosphoric 'Envelope': Scientific Appraisal, Domestic Spectacle, and (Un)'Reasonable Creatures' in Edgeworth's *Belinda*." *Eighteenth-Century Fiction* 24, no. 3 (Spring 2012): 509–36.

Zytaruk, Maria. "Mary Delany." In Laird and Weisberg-Roberts, eds., *Mrs. Delany and Her Circle*, 130–49.

INDEX